Network Reliability
and
Algebraic Structures

Network Reliability and Algebraic Structures

DOUGLAS R. SHIER
College of William and Mary,
Williamsburg, Virginia

CLARENDON PRESS · OXFORD

This book has been printed digitally and produced in a standard specification in order to ensure its continuing availability

OXFORD
UNIVERSITY PRESS

Great Clarendon Street, Oxford OX2 6DP
Oxford University Press is a department of the University of Oxford.
It furthers the University's objective of excellence in research, scholarship,
and education by publishing worldwide in

Oxford New York

Auckland Cape Town Dar es Salaam Hong Kong Karachi
Kuala Lumpur Madrid Melbourne Mexico City Nairobi
New Delhi Shanghai Taipei Toronto
With offices in
Argentina Austria Brazil Chile Czech Republic France Greece
Guatemala Hungary Italy Japan South Korea Poland Portugal
Singapore Switzerland Thailand Turkey Ukraine Vietnam

Oxford is a registered trade mark of Oxford University Press
in the UK and in certain other countries

Published in the United States
by Oxford University Press Inc., New York

© D. R. Shier, 1991

The moral rights of the author have been asserted

Database right Oxford University Press (maker)

Reprinted 2011

All rights reserved. No part of this publication may be reproduced,
stored in a retrieval system, or transmitted, in any form or by any means,
without the prior permission in writing of Oxford University Press,
or as expressly permitted by law, or under terms agreed with the appropriate
reprographics rights organization. Enquiries concerning reproduction
outside the scope of the above should be sent to the Rights Department,
Oxford University Press, at the address above

You must not circulate this book in any other binding or cover
And you must impose this same condition on any acquirer

ISBN 978-0-19-853386-3

Preface

The study of reliability, in particular the reliability of networks, has received considerable impetus from its clear applicability to computer networks, communication systems, and distribution systems. Moreover, there has been increased interest in the subject because of the variety of discrete, combinatorial, and algebraic mathematics that can be found lurking just underneath this practical veneer. As might be expected, not only can existing mathematical tools be used to analyse the reliability of networks, but new approaches developed in the context of network reliability can have ramifications beyond that particular venue. Rather than attempting to address network reliability and its many mathematical underpinnings, this monograph attempts to weave a closely knit story by emphasizing a single mathematical strand, that of algebraic structures underlying network reliability problems. In the chapters that follow, various types of algebraic objects (partially ordered sets, lattices, semilattices, and spaces of polynomials) emerge to play their roles in simplifying, unifying, and comprehending the nature of network reliability problems.

The entire monograph is meant to be fairly self-contained, with little mathematical background assumed of the reader. Of course, familiarity with the basic notions of partial orders and elementary graph theory will be useful in appreciating the subject. Pertinent concepts enlisted from these areas of mathematics are developed as they arise in the discussion of network reliability. In addition, a glossary of technical terms peculiar to these areas is found at the end of the last chapter, and this should aid the reader who is uncertain of the precise meaning of particular terms. Notes at the end of each chapter direct attention to various sources for material presented in the chapter, while also pointing to additional information on related topics.

The first chapter provides a brief survey of the entire area of network reliability, giving an overview of various network models (deterministic and probabilistic) as well as a selection of important vulnerability and reliability measures. This chapter serves to establish the broader perspective into which the material of this monograph fits, and it also affords the opportunity to introduce certain basic terminology. The remainder of the monograph initially focuses on a specific probabilistic network model, the two-terminal reliability problem in directed networks, although there are occasional references to related results for other variations on this specific theme. Chapter 2 discusses a variety of general approaches for calculating network reliability: state-space enumeration, inclusion–exclusion, disjoint products, and factoring. It turns out that each of these methods can either be analysed or fine-tuned using an algebraic approach, and so later chapters return to these general solution approaches with additional algebraic tools brought to bear. The final section of the chapter illustrates why the computational requirements of these algorithmic approaches in general explode with increasing problem size, thus providing an informal introduction to the notions of NP-completeness and #P-completeness.

Chapter 3 initiates an algebraic perspective, applied to the two-terminal reliability problem. Rather than simply seeking a numerical answer for network reliability, one instead insists upon a functional representation, as a polynomial in the given edge reliabilities. This additional structure permits symbolic (rather than numeric) manipulation using suitably defined algebraic operations, and the reliability polynomial then assumes a pleasingly concise form when expressed in terms of these operations. Given this algebraic representation, determining the reliability polynomial can be viewed as solving a system of 'linear' equations in these new operations. As a result, the considerable apparatus of numerical linear algebra can be invoked to yield algebraic methods for finding the reliability polynomial. In particular, various iterative schemes are derived and analysed for the two-terminal network reliability problem.

Chapter 4 considers a number of methods for the approximate, rather than the exact, calculation of two-terminal reliability. Several techniques for producing lower and upper bounds on network reliability are surveyed. Algebraic methods are also developed for producing a sequence of lower- and upper-bounding polynomials for the reliability polynomial: these polynomials do indeed bound, in an algebraic sense, the reliability polynomial. Moreover, such bounding polynomials are guaranteed to converge, in a finite number of iterations, to the exact reliability polynomial. However, for obvious computational reasons, the process can be terminated after relatively few iterations, yielding absolute lower and upper bounds on the unknown reliability value.

Various exact and approximate methods for calculating network reliability require the prior enumeration of the paths or the cutsets of a network. Consequently, Chapter 5 discusses several techniques for generating, in a relatively efficient manner, the paths and cutsets of an arbitrary network. The underlying algebraic structure of Chapter 3 is modified so that these objects can also be enumerated by purely algebraic methods. Chapter 6 makes use of such generated objects to produce 'pseudopolynomial' algorithms for calculating network reliability; that is, the computational complexity of such algorithms is polynomially bounded in the number of such objects (path or cutsets). The approach is phrased within the more general framework of coherent systems, and system reliability is calculated using a certain semilattice to guide the disjoint-products technique. The application of Möbius inversion to this semilattice also reveals an additional perspective on network reliability calculations.

Chapter 7 discusses a different type of reliability problem, that of stochastically covering a set with certain of its subsets. Solution of this reliability covering problem is shown to be computationally equivalent to the calculation of coherent system reliability, demonstrating that the covering problem is also in general provably hard. However, two special cases, involving the covering of nodes in a tree, are analysed and shown to be polynomially solvable.

Chapter 8 considers again the technique of state-space enumeration, but now invoked as an approximation technique. In fact, this method is generally applicable to any system of failure-prone components, not necessarily those defined by the elements of a network. Implementation of state-space approximation naturally leads to the problem of generating the states of such a system in order of nonincreasing probability. This can be carried out, in a relatively efficient manner, by identifying an underlying algebraic structure (a distributive lattice). Examination of the Hasse diagram of this distributive lattice leads in turn to an algorithm for generating the most probable states of the system, without the need to consult the entire state space. The performance of a coherent system in newly generated states can be flawlessly predicted from the performance in previously generated states, by first developing the concept of critical sets. An extension of the problem of generating most probable states in 'multistate' systems is also considered in the chapter.

The algebraic perspective enunciated in this monograph ties together several reliability models and solution techniques. It not only provides a unifying framework for viewing existing results but it also suggests new directions for the application of discrete mathematical tools. Of course, this approach represents but one facet of the renewed mathematical interest shown in network reliability problems. For example, there are a number of fascinating and powerful combinatorial approaches to the study of network reliability that do not fall within the compass of this present work. Fortunately, the recent monograph by Charles Colbourn (*The combinatorics of network reliability*, Oxford University Press, 1987) admirably distills and presents the essence of this complementary combinatorial thread.

The development of the present monograph has been greatly aided and accelerated by two separate opportunities to present portions of this work. The first occurred at the Rocky Mountain Mathematics Consortium Regional Summer School, sponsored by the National Science

Foundation and held at the University of Wyoming during the summer of 1989. The second opportunity arose as a Distinguished Visiting Professor in the Mathematical Sciences Department at Clemson University during a week-long visit there during the fall of 1989. Since 1984, the United States Air Force Office of Scientific Research has provided continual support and encouragement for the research carried out on algebraic aspects of network reliability problems. The author also acknowledges the support offered by the Applied and Computational Mathematics Division of the National Institute of Standards and Technology.

D.R.S.

Williamsburg, Virginia
October, 1990

Contents

1. Overview of network reliability 1
   1.1 Network models 2
   1.2 Performance measures 3
   1.3 Chapter notes 7

2. Approaches for calculating network reliability 8
   2.1 State-space enumeration 8
   2.2 Inclusion–exclusion 10
   2.3 Disjoint products 13
   2.4 Factoring 15
   2.5 Computational complexity 18
   2.6 Chapter notes 20

3. An algebraic formulation of network reliability
   problems 22
   3.1 Algebraic operations 23
   3.2 Linear equations 25
   3.3 Iterative methods 29
   3.4 Computational aspects 33
   3.5 Chapter notes 39

4. Bounds on two-terminal reliability 41
   4.1 Bounding techniques 41
   4.2 Algebraic bounds 48
   4.3 Computation of algebraic bounds 51
   4.4 Chapter notes 60

5. Enumeration of paths and cutsets 62
   5.1 Path-enumeration methods 62
   5.2 Cutset-enumeration methods 67
   5.3 Chapter notes 71

6. Pseudopolynomial algorithms for calculating reliability — 72
   6.1 Examples of lattice structures — 72
   6.2 A recursive algorithm for semilattices — 75
   6.3 Möbius inversion and reliability calculations — 80
   6.4 Chapter notes — 83

7. Reliability covering problems — 84
   7.1 A motivating example — 84
   7.2 The complexity of covering problems — 87
   7.3 An algorithm for undirected trees — 91
   7.4 An algorithm for directed trees — 95
   7.5 Refinements and extensions — 98
   7.6 Chapter notes — 100

8. State-space approximation — 101
   8.1 Lattice structure — 102
   8.2 An algorithm for state generation — 106
   8.3 Critical sets — 110
   8.4 Multistate systems — 113
   8.5 Chapter notes — 117

9. Stochastic shortest-path problems — 118
   9.1 Algebraic structure — 119
   9.2 Fundamental reductions — 122
   9.3 Conditioned reductions — 124
   9.4 Computational results — 126
   9.5 Chapter notes — 129

Glossary of terms — 130

References — 133

Index — 143

1
Overview of network reliability

The broad theme of reliability theory is the study of the overall performance of a system comprising failure-prone elements. That is, the components of the system are not perfect in operation, but their failure is assumed to be governed by certain probabilistic laws. It is thus of interest to describe the stochastic behavior of the system in terms of the given stochastic characteristics of its components. Not only are the reliabilities of individual components of importance, but the manner in which they are assembled can have a significant influence on the overall performance of the system. For example, as noted over thirty years ago by Moore and Shannon (1956), it is possible to obtain a reliable system by properly configuring unreliable components through the use of redundancy.

The problem of determining the reliability of a complex system, whose components are subject to failure, has received considerable attention in the statistical, engineering, and operations research literature. Indeed, in certain situations, improving the reliability of a system can be more consequential than reducing its cost. Reliability analysis can be applied to a variety of practical systems, ranging from large-scale telecommunication, transportation, and mechanical systems, to the microelectronic scale of integrated circuits. In addition, the reliability of computer software, as well as computer hardware, has become increasingly important in commercial and military applications. In examples such as these, there are two broad objectives addressed by reliability theory: assessing the reliability of a given system (analysis) and designing as reliable a system as possible from the given components (synthesis).

We will be exclusively concerned here with network reliability, in which the underlying system arises from the interconnection of various elements in the form of a *network,* or *graph,* such as is exemplified by telecommunication, distribution, and computer networks. For example, the *nodes* of a computer communication network might represent the physical locations of computers and its *edges* might represent existing communication links between computer sites. Alternatively, the network might represent an electrical power system, connecting users (nodes) to one another and to the power grid by various transmission lines (edges). In realistic settings, the elements of a network, its nodes or edges or both, are subject to failure. At any instant, each element is either working or failed, and, as a result, the network itself is also either working or failed. In the computer communication example, working might mean that a distinguished sender and distinguished receiver are able to communicate over operational links of the network, while failure means that no complete transmission path is available. In the power system example, the network might be considered to have failed if any users are unable to derive power from the system. As can be seen, there are a variety of situations that can be captured through appropriate measures of network reliability. In the following sections, we will provide a brief overview of the diversity of network reliability problems, classified according to the underlying network model and the selected performance measure.

1.1 Network models

A network $G = (N, E)$ consists of a set N of nodes together with a set E of edges, representing pairs of nodes. If the pairs are ordered, then we have a *directed* network and (i, j) represents the directed edge joining node i to node j. If the pairs are considered to be unordered, then we have an *undirected* network and the edge joining i and j is represented by $[i, j]$. In either case the edge between i and j is said to be *incident* to both i and j; in this case nodes i and j are said to be *adjacent*. For most purposes an undirected network can be adequately represented as a directed network by associating oppositely directed edges (i, j) and (j, i) with each undirected edge $[i, j]$.

At any instant the elements of the network (nodes and/or edges) will be in either of two possible states, working or failed. In *deterministic* networks it is considered that working elements can be successfully attacked by an adversary, resulting in their failure or inactivation. The failure of an edge means that it is removed from the network, while the failure of a node means that the node and all its incident edges are removed from the network. In deterministic network models the focus is typically on evaluating the worst-case performance of the network, in which the adversary intelligently chooses certain elements to inactivate, resulting in the maximum damage to the network. This type of model thus provides a conservative assessment of performance, and it would be particularly appropriate in the design of robust military systems.

By contrast, in *probabilistic* networks it is usually assumed that, at any instant, elements fail randomly and independently of one another, according to certain known probabilities. Specifically, each node i has an associated *reliability* p_i indicating the probability that it is operational, and each edge k has a reliability p_k, the probability that it is operational. Thus at any instant (or 'snapshot') the elements of the network fail independently with probabilities $q_i = 1 - p_i$ and $q_k = 1 - p_k$, respectively.

In these circumstances, one would be interested in assessing the average performance of the network, under the assumption of random (as opposed to malevolent) failures. It is also possible to allow for dependent failure modes, at the expense of added data-gathering requirements and increased subsequent computation. For example, the edges incident with a given node might be subject to certain common influences (such as weather, interference, or jamming), and these edges might therefore tend to fail together, rather than independently; or the failure of one edge might place additional stress on the other operating incident edges, making them more likely to fail.

1.2 Performance measures

Before discussing various performance measures for deterministic and probabilistic networks, a few preliminary definitions are required. For ease of exposition it will be assumed that the network G is undirected, since the concepts easily extend to the directed case. The *degree* of a node is the number of edges incident with the node. An *i–j path* of *length* k in the network $G = (N, E)$ is a sequence of nodes $i = u_0, u_1, \ldots, u_k = j$ with $(u_{r-1}, u_r) \in E$ for all $r = 1, \ldots, k$. If all nodes on the path (except possibly $i = j$) are distinct, the path is termed *simple*. The *distance* between nodes i and j is the length of a shortest path joining them, and the *diameter* of the network is the maximum distance achieved between nodes of G. If there is a path joining every pair of distinct nodes, then G is *connected*. An *edge disconnecting set* of a connected network G is a set of edges whose removal disconnects G. In a similar way, a *node disconnecting set* of a connected network G is a set of nodes whose removal disconnects G or leaves the trivial network (a single node). Recall that the removal of a node also eliminates any edges incident with the node. A *cut node* is a node whose removal from G disconnects the network.

A number of *vulnerability* measures are available for assessing the worst-case behavior of a deterministic network. In order to measure how far a network is from being disconnected, several connectivity-based measures have been proposed. One such measure is the *edge connectivity* $\lambda(G)$ of the network, the minimum cardinality of an edge disconnecting set. This measure would be especially meaningful when connecting all nodes in the network is of primary importance. In the case where communication between two specified nodes s and t is of particular importance, the *s-t edge connectivity* $\lambda_{st}(G)$ would be relevant: that is, the minimum cardinality of an edge disconnecting set that separates s from t. Clearly, $\lambda(G)$ equals the minimum value of $\lambda_{st}(G)$ over all s, $t \in N$. Other measures concentrate instead on node failures rather than edge failures. The *node connectivity* $\kappa(G)$ is the minimum cardinality of a node disconnecting set. Also, the *s-t node connectivity* $\kappa_{st}(G)$ is the minimum cardinality of a node disconnecting set that separates s from t, where s and t are not adjacent.

Other things being equal, one would prefer a network with as large a connectivity as possible. The following well-known result (Harary 1969) shows that there are simple bounds on how large these various parameters can be in a network G composed of n nodes and m edges:

$$\kappa(G) \leq \lambda(G) \leq \delta(G) \leq \lfloor 2m/n \rfloor,$$

where $\delta(G)$ indicates the minimum degree of a node in G, and $\lfloor x \rfloor$ is the greatest integer less than or equal to x. Thus, for a given number of nodes and edges, it is necessary to make the degrees as equal as possible in order to maximize $\kappa(G)$ and $\lambda(G)$. Even so, two networks can have the same (maximum) connectivity but one can be more readily disconnected than the other, since there might be more ways of disconnecting the first network than the second. This idea suggests the relevance of other vulnerability measures, such as μ_k, the number of edge disconnecting sets of size k in G. Notice that $k = \lambda(G)$ is the smallest k for which $\mu_k > 0$. The vector $\mu = (\mu_1, \ldots, \mu_m)$ of these quantities is called the *cut frequency vector* of the network (Boesch 1985); a similar such vector can be defined relative to node disconnecting sets.

An alternative measure, called the *cohesion* $\mu(G)$ of a network G, measures the minimum number of edges whose removal creates a cut node in the network (Ringeisen and Lipman 1983). Thus removal of these $\mu(G)$ edges, together with a single node, will disconnect the network. Other types of mixed measures (involving the removal of both nodes and edges) have been proposed. For example, the *connectivity function* (Beineke and Harary 1967) specifies for each i the minimum edge connectivity that can be achieved after first removing i nodes.

When a network becomes disconnected it is desirable to capture the extent of disruption by measuring the size and number of the remaining *connected components*. After all, a system which has been split into many small parts represents a more severe disruption than one which has been split into a few large parts. Several parameters have been studied for combining the size of the disconnecting set with the characteristics of the components that remain. To aid in describing such measures suppose edges $S \subseteq E$ are removed from G yielding the network $G - S$ with $c(G - S)$ connected components and maximum component size $m(G - S)$. Then the *ith-order edge connectivity* (Boesch and Chen 1978) is defined as the minimum $|S|$ such that $c(G - S) = i + 1$, while the *ith edge separation value* (Beineke et al. 1990) is defined as the minimum $|S|$ such that $m(G - S) \leq i$. Instead of producing a sequence of values in this way, we can distill an overall measure reflecting the nature of the elements that fail and the structure that remains intact. The *edge integrity* (Barefoot *et al.* 1987) of G is the minimum value of the sum $\{|S| + m(G - S)\}$ over all $S \subseteq E$, while the *edge toughness* of G (Chvátal 1973) is the minimum value of the ratio $\{|S|/c(G - S)\}$ over all disconnecting sets S. Ideally one would design networks so that these measures are large. It turns out that the edge toughness of G always equals $\frac{1}{2}\lambda(G)$, so a more interesting measure is the *node toughness*: the minimum ratio of the size of a node disconnecting set to the resulting number of components (Chvátal 1973). Indeed, most of the edge-based vulnerability measures discussed here have appropriate analogues in the node failure case.

Additional measures of performance in a deterministic network G can be based on distances between nodes in the network. For example, the diameter of G indicates the maximum time required for a message exchange between any two nodes in the system, assuming all edges have the same transmission time. It would be important to design a network not only with a reasonably small diameter but such that the failure of any edge increases the diameter by as little as possible. Another measure related to distance is the *persistence* of the network (Boesch *et al.* 1981), the minimum number of edges that must be removed in order to increase the diameter or disconnect G.

For the case of probabilistic networks (in which nodes and/or edges fail randomly and independently with known probabilities), a number of measures have been explored. Suppose that G is directed, with s and t being distinguished nodes of G. The *two-terminal reliability* $R_{st}(G)$ is the probability that s and t are connected by a path of operating edges and nodes in G. The *source-to-all-terminal reliability* $R(G)$ is the probability that there is an operative path from node s to all other nodes of the network. The all-terminal reliability would be an appropriate measure when all nodes are of equal importance in receiving a message sent from the source node, whereas the two-terminal reliability would apply when a critical message needs to be routed between specified sites in the network. A generalization of these concepts is embodied in the *source-to-K-terminal reliability* of the network: the probability $R_K(G)$ that there is an operative path from node s to all nodes in some specified set $K \subseteq N$. These probabilistic measures have analogous counterparts in the case where G is undirected. Notice that for undirected networks the all-terminal reliability $R(G)$ simply expresses the probability that G remains connected.

An alternative probabilistic measure takes into account the fact that different ways of disconnecting a network are of different severity. For example, if G is undirected and connected, then the failure of certain edges and nodes could separate G into several connected components

$G_1, \ldots, G_r$. All communication is then disrupted between nodes in different components, and the resulting communication capacity can be measured by the number of pairs of nodes now able to communicate:

$$\sum_{i=1}^{r} \binom{n_i}{2},$$

where n_i is the number of nodes in component G_i. The average number of node pairs able to communicate, taken over all possible node and edge failures, thus provides a quite different type of probabilistic measure, the *pair-connectivity* or *resilience* of G (Amin et al. 1988; Colbourn 1987b). Cohen (1986) has recently studied a measure of *redundancy* in networks, namely the average number of ways the system can remain connected. Rather surprisingly, there is a polynomial-time algorithm for calculating this stochastic measure, using a determinantal formula.

This brief survey of vulnerability and reliability measures illustrates the diverse nature of network reliability problems. Given a network model and some performance measures of interest there naturally arise the issues of both analysis and synthesis. That is, effective computational procedures are first examined for evaluating the selected measures. Then the synthesis phase would attempt to construct a network, chosen from a class of networks with prescribed characteristics, having the best performance possible with respect to such measures. For instance, one might seek a deterministic network on a given number of nodes and edges having maximum connectivity and yet acceptably small diameter. Alternatively, in the probabilistic case, one might desire a network on a given number of nodes and edges having maximum two-terminal reliability relative to the given node and edge failure probabilities. It should be clear that changing these input probabilities might very well change the structure of the 'optimum' network. Recent evidence has indeed shown this to be the case for many probabilistic reliability measures, except in special circumstances (Amin et al. 1988; Myrvold 1989).

Numerous algorithms have been developed for calculating the various network performance measures discussed here. While the calculation of deterministic measures can often be carried out in an efficient manner, the effective computation of virtually all probabilistic measures for general networks remains elusive. Namely, all known general procedures exhibit a worst-case behavior that is exponential in the size of the network. The reason for this will become clear in the discussion of computational complexity in Chapter 2, which also presents an overview of the most popular techniques for analysing probabilistic networks. In view of the notorious difficulty of probabilistic network problems, the exact computation of network reliability has been confined to networks of rather small size. As a result, recent research activity has concentrated on developing algorithms adapted to special network structures (where polynomial-time algorithms are possible) or has pursued approximation techniques applicable to general networks.

1.3 Chapter notes

This chapter has provided a brief introduction to a variety of models and performance measures for assessing the reliability (or availability) of networks. An historical overview of the evolution of reliability theory, culminating in the recent emphasis on network reliability, is presented by Barlow (1984). Surveys of the progress to date (through the early 1970s) on methods for the analysis and synthesis of deterministic networks are found in the papers of Frank and Frisch (1970) and Wilkov (1972), as well as in the book by Frank and Frisch (1971). These sources also treat, in a more limited way, the analysis and synthesis of probabilistic networks. A more recent survey, focusing on the design of reliable networks with respect to a variety of deterministic criteria, is presented by Boesch (1986).

Additional information on a number of vulnerability measures and their calculation for certain network structures are to be found in the recent papers of Bagga *et al.* (1989), Barefoot *et al.* (1987), and Beineke (1989). Other connectivity measures are discussed in the work of Lipman and Pippert (Lipman and Pippert 1985; Pippert and Lipman 1985). Peyrat (1984) studies certain vulnerability measures associated with distance (and connectivity) in networks. Satyanarayana (1982) discusses and develops a number of reliability measures for the analysis of probabilistic networks.

Future directions of reliability analysis in the areas of software reliability, fault-tolerant systems, and Bayesian approaches are summarized in the paper of Barlow and Singpurwalla (1985). A review of network reliability models, and their applicability to computer communication systems, is provided in the paper of Spragins *et al.* (1986). More realistic extensions to current models, such as the incorporation of dependent failures and system degradation, are also emphasized in this paper.

2

Approaches for calculating network reliability

This chapter provides a survey of several general approaches for calculating the reliability of probabilistic networks. Throughout the remainder of this book, a specific network model and performance measure will be emphasized. Namely, it is supposed that $G = (N, E)$ is a directed network, having a distinguished source node s and distinguished destination node t. The nodes of G are assumed to be perfect, whereas the edges $k \in E$ are assumed to fail in a statistically independent fashion with known probabilities $q_k = 1 - p_k$. Of particular concern here will be the two-terminal reliability measure $R_{st}(G)$, the probability that there is a path of operative edges from s to t in G. This paradigm not only serves to illustrate the essential features of the approaches discussed in this chapter, but it also features prominently in the algebraic methods that form the core of the remaining chapters.

2.1 State-space enumeration

The most fundamental method of calculating $R_{st}(G)$ uses state-space enumeration and dates back to Moore and Shannon (1956). Since each of the $m = |E|$ edges of G assumes one of two states, working or failed, the state of the network can be represented using a 0-1 vector $\delta = (\delta_1, \delta_2, \ldots, \delta_m)$. The kth component of δ equals 1 if edge k is working and is 0 otherwise. Under the assumption of independent edge failures, the probability of a given state δ is

$$\Pr(\delta) = \prod_{k=1}^{m} p_k^{\delta_k}(1-p_k)^{1-\delta_k}.$$

Define the 0-1 variable $I_{st}(\delta)$, which equals 1 precisely when the subnetwork of operational edges k (having $\delta_k = 1$) contains an s-t path. Then the two-terminal reliability is given by

$$R_{st}(G) = \sum_{\delta \in \mathcal{D}} I_{st}(\delta) \Pr(\delta), \tag{2.1}$$

where $\mathcal{D}$ is the set of all network states. Although conceptually simple, the state-space approach is impractical because $|\mathcal{D}| = 2^m$. As an illustration, consider the network on four nodes and five edges shown in Fig. 2.1. It is clear that the system works (i.e., contains an s-t path) if at

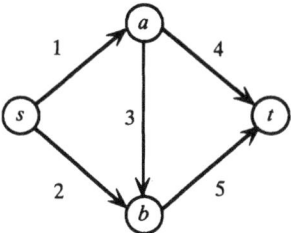

Fig. 2.1 An illustrative network.

most one edge fails, or if any two edges other than $\{1,2\}$, $\{1,5\}$ or $\{4,5\}$ fail. On the other hand, for three or more edge failures, the system fails unless the failed edges are $\{1,3,4\}$ or $\{2,3,5\}$. Consequently, the two-terminal reliability of this network is given by

$$R_{st}(G) = p_1 p_2 p_3 p_4 p_5 + q_1 p_2 p_3 p_4 p_5 + p_1 q_2 p_3 p_4 p_5 + p_1 p_2 q_3 p_4 p_5 +$$
$$p_1 p_2 p_3 q_4 p_5 + p_1 p_2 p_3 p_4 q_5 + q_1 p_2 q_3 p_4 p_5 + q_1 p_2 p_3 q_4 p_5 +$$
$$p_1 q_2 q_3 p_4 p_5 + p_1 q_2 p_3 q_4 p_5 + p_1 q_2 p_3 p_4 q_5 + p_1 p_2 q_3 q_4 p_5 +$$
$$p_1 p_2 q_3 p_4 q_5 + q_1 p_2 q_3 q_4 p_5 + p_1 q_2 q_3 p_4 q_5.$$

Substitution of $q_k = 1 - p_k$ into the above produces, after some algebraic simplification,

$$R_{st}(G) = p_1 p_2 p_3 p_4 p_5 - p_1 p_2 p_3 p_5 - p_1 p_2 p_4 p_5 - p_1 p_3 p_4 p_5 + p_1 p_3 p_5 +$$
$$p_1 p_4 + p_2 p_5.$$

Although as many as 55 terms could have resulted from performing these substitutions, a good deal of cancellation occurred in producing the above expression. Moreover, the coefficients appearing in the simplified form are either $+1$ or -1. This phenomenon is no accident, as will be seen in the next section.

Since only states $\boldsymbol{\delta}$ with $I_{st}(\boldsymbol{\delta}) = 1$ contribute to the sum (2.1), it is unnecessary to examine all states of $\mathcal{D}$, merely those containing an s-t path. It is therefore appropriate to focus directly on the simple s-t paths $\{P_1, P_2, \ldots, P_k\}$ of G. Define E_i to be the event that all edges in path P_i operate. Then the two-terminal reliability is the probability that at least one such event occurs, or

$$R_{st}(G) = \Pr(E_1 \cup E_2 \cup \cdots \cup E_k). \qquad (2.2)$$

Alternatively, the two-terminal reliability can be formulated using the minimal s-t edge disconnecting sets, or *cutsets*, of G; an s-t edge disconnecting set is minimal if it does not contain any other edge disconnecting set separating s and t. Indeed, suppose that the s-t cutsets of G are $\{C_1, C_2, \ldots, C_r\}$ and let F_j be the event that all edges in cutset C_j fail. Then the two-terminal *unreliability* $U_{st}(G)$ is given by

$$U_{st}(G) = 1 - R_{st}(G) = \Pr(F_1 \cup F_2 \cup \cdots \cup F_r). \tag{2.3}$$

Unfortunately, the events E_i appearing in (2.2) are not in general disjoint, nor are the events F_j appearing in (2.3). There are, however, standard methods for evaluating the probability of the union of events, and the following two sections discuss their application to network reliability problems.

Another way of viewing state-space enumeration emerges from the binary nature of the states assumed by each edge. Rather than fully specifying the states of all m edges at once, we can instead select a particular edge $e \in E$ and 'condition' on the status of e, either perfect ($p_e = 1$) or failed ($p_e = 0$). In the first case, we obtain a new system denoted G/e in which edge e is perfect, and, in the second, a new system $G - e$ in which e is failed. This produces the *pivotal decomposition formula* (Moskowitz 1958):

$$R_{st}(G) = p_e R_{st}(G/e) + (1 - p_e)R_{st}(G - e). \tag{2.4}$$

This formula shows how reliability calculations for a given system can be decomposed into those for two smaller systems, G/e and $G - e$. While conditioning, or *factoring*, in turn on every possible edge just reproduces state-space enumeration, there are circumstances in which not all edges need to be considered for factoring. In fact, by judiciously selecting the edges for factoring (see Section 2.4), substantial computational savings can be achieved.

2.2 Inclusion–exclusion

Expression (2.2) can be expanded using the principle of inclusion and exclusion, yielding

$$R_{st}(G) = \sum_i \Pr(E_i) - \sum_{i<j} \Pr(E_i E_j) + \sum_{i<j<l} \Pr(E_i E_j E_l) - \cdots$$
$$+ (-1)^{k+1} \Pr(E_1 E_2 \cdots E_k).$$

Here the intersection of events A and B is indicated by the juxtaposition AB. By the independence assumption, each term in this expansion is easy to calculate. However, there are $2^k - 1$ terms appearing, exponential in the number of given paths. For the example of Fig. 2.2, the 'bridge network,' there are four simple s-t paths

$$P_1: \text{1-5}, \quad P_2: \text{2-6}, \quad P_3: \text{1-3-6}, \quad P_4: \text{2-4-5}.$$

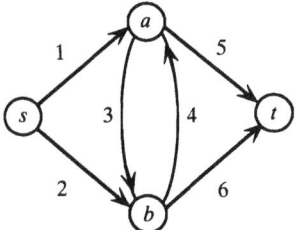

Fig. 2.2 The bridge network.

As a result, $\Pr(E_1) = p_1 p_5$, $\Pr(E_1 E_2) = p_1 p_2 p_5 p_6$, $\Pr(E_1 E_3) = p_1 p_3 p_5 p_6$, $\Pr(E_1 E_2 E_3) = p_1 p_2 p_3 p_5 p_6$, and so forth. Application of the inclusion–exclusion method then produces the expression

$$\begin{aligned}R_{st}(G) = &(p_1 p_5 + p_2 p_6 + p_1 p_3 p_6 + p_2 p_4 p_5) - \\&(p_1 p_2 p_5 p_6 + p_1 p_3 p_5 p_6 + p_1 p_2 p_4 p_5 + p_1 p_2 p_3 p_6 + \\&p_2 p_4 p_5 p_6 + p_1 p_2 p_3 p_4 p_5 p_6) + \\&(p_1 p_2 p_3 p_5 p_6 + p_1 p_2 p_4 p_5 p_6 + 2 p_1 p_2 p_3 p_4 p_5 p_6) - \\&(p_1 p_2 p_3 p_4 p_5 p_6) \\= &p_1 p_5 + p_2 p_6 + p_1 p_3 p_6 + p_2 p_4 p_5 - p_1 p_2 p_5 p_6 - p_1 p_3 p_5 p_6 - \\&p_1 p_2 p_4 p_5 - p_1 p_2 p_3 p_6 - p_2 p_4 p_5 p_6 + p_1 p_2 p_3 p_5 p_6 + p_1 p_2 p_4 p_5 p_6.\end{aligned}$$

Rather than the 15 possible terms appearing, cancellation has resulted in only 11 terms. In addition the coefficients of terms remaining in the reduced expression are either $+1$ or -1. This remarkable phenomenon was first studied by Satyanarayana and Prabhakar (1978), who provided an explanation for this simplification and presented an algorithm based on the special form of the reduced inclusion–exclusion expression.

In order to understand the nature of noncancelling terms and their sign, first notice that direct expansion of the inclusion–exclusion formula generates terms corresponding to certain sets of edges. In the above example, the term $p_1 p_2 p_4 p_5$ corresponds to a particular subnetwork H having node set $N(H) = \{s, a, b, t\}$ and edge set $E(H) = \{1, 2, 4, 5\}$. More generally, $R_{st}(G)$ can be expressed as a sum over all possible subnetworks H of G

$$R_{st}(G) = \sum_H d(H) \Pr(H), \qquad (2.5)$$

where $\Pr(H) = \prod \{p_e : e \in E(H)\}$ is, by independence, the probability that all edges in H operate and $d(H)$ is its associated coefficient. Because the terms of the inclusion–exclusion formula arise by superimposing s-t

paths, certain subnetworks are guaranteed to have a coefficient of 0. Namely, if H contains an *irrelevant* edge, one not on any simple s-t path, then $d(H) = 0$. The following result (Satyanarayana and Prabhakar 1978) shows that any subnetwork H containing a *cycle* (a closed directed path) likewise must have $d(H) = 0$. Moreover, acyclic subnetworks H containing no irrelevant edges will produce a coefficient $d(H)$ of ± 1.

Theorem 2.1. *Suppose that G is a directed network with distinguished nodes s and t. In the reduced inclusion–exclusion expansion for $R_{st}(G)$, $d(H) = 0$ for any subnetwork H containing a cycle and $d(H) = (-1)^{E(H)-N(H)+1}$ for any acyclic subnetwork H having no irrelevant edges.*

Proof. The essential idea is to use the pivotal decomposition formula applied to an edge e directed out of node s, and then to proceed by induction on the number of edges m. By equating coefficients of $p_1 p_2 \cdots p_m$ in (2.4), it follows that

$$d(G) = d(G/e) - d(G - e). \tag{2.6}$$

Because e is an edge out of the source node, G/e corresponds to the network obtained from G by removing edge e and contracting its endpoints. The system $G - e$ corresponds to removing edge e from G. In what follows, we can assume that H has at least three nodes and contains no irrelevant edges; in particular H can have no edges leading into node s. If H is acyclic, its nodes can be numbered so that every directed edge (i, j) satisfies $i < j$; we may then assume $s = 1$. By selecting edge $e = (s, j)$ with j having the smallest such number, it is assured that node j has exactly one edge entering it, namely e. Since e is relevant, there will exist some edge f leaving node j, so that, in $H - e$, edge f will be irrelevant and $d(H - e) = 0$, whence $d(H) = d(H/e)$. Since H/e remains acyclic with no irrelevant edges and since $E(H/e) = E(H) - 1$, $N(H/e) = N(H) - 1$, application of induction now yields $d(H) = d(H/e) = (-1)^{[E(H)-1]-[N(H)-1]+1} = (-1)^{E(H)-N(H)+1}$. Suppose, on the other hand, that H contains a cycle but no irrelevant edges. Then select any edge e out of node s and apply the pivotal decomposition formula. Edge e cannot be part of a cycle (there are no edges in H entering s) so that $H - e$ still contains a cycle. Also, H/e will contain a cycle, whence, by induction, $d(H) = d(H/e) - d(H - e) = 0$. □

The importance of Theorem 2.1 is that only noncancelling terms of the inclusion–exclusion formula are now represented and the sign of each noncancelling term is predictable. For example, the directed network of Fig. 2.2 contains the acyclic subnetwork H with $E(H) = \{1, 2, 3, 5, 6\}$ and $N(H) = N$ so that $p_1 p_2 p_3 p_5 p_6$ will appear with a sign of $(-1)^{5-4+1} = +1$; however, the term $p_1 p_2 p_3 p_4 p_5 p_6$ will not appear since the corresponding subnetwork contains a cycle. In general, the calculation of $R_{st}(G)$ can be reduced to generating all acyclic subnetworks having relevant edges. In this sense the topological formula of Satyanarayana and Prabhakar is the most efficient method based on the inclusion–exclusion approach, although the number of terms in the reduced expression can still grow rapidly with the problem size. A reduced inclusion–exclusion formula for $R_K(G)$, governed by an analogue of Theorem 2.1, holds in directed networks (Satyanarayana 1982); it can be established by modifying the proof given here. Boesch *et al.* (1990) discuss various combinatorial interpretations of the formula for $R_K(G)$.

2.3 Disjoint products

Another way to calculate the probability of the union of events in (2.2) is to decompose $E_1 \cup E_2 \cup \cdots \cup E_k$ into a union of events that are disjoint. Specifically, we can express

$$R_{st}(G) = \Pr(E_1 \cup E_2 \cup \cdots \cup E_k)$$
$$= \Pr(E_1 \cup \bar{E}_1 E_2 \cup \bar{E}_1 \bar{E}_2 E_3 \cup \cdots \cup \bar{E}_1 \bar{E}_2 \bar{E}_3 \cdots \bar{E}_{k-1} E_k),$$

where $\bar{E}_i$ denotes the complement of event E_i. Since the compound events above are pairwise disjoint,

$$R_{st}(G) = \Pr(E_1) + \Pr(\bar{E}_1 E_2) + \Pr(\bar{E}_1 \bar{E}_2 E_3) + \cdots$$
$$+ \Pr(\bar{E}_1 \bar{E}_2 \bar{E}_3 \cdots \bar{E}_{k-1} E_k).$$

This *disjoint-products* method involves adding only k probabilities; however, the calculation of each constituent probability is, in general, quite involved. It is also important to emphasize that the efficacy of this method can be highly dependent on the specific ordering given to the events E_i.

As an illustration, consider the calculation of two-terminal reliability for the network of Fig. 2.2, in which the paths are presented in the order

$$P_1: \ 1\text{-}5, \qquad P_2: \ 1\text{-}3\text{-}6, \qquad P_3: \ 2\text{-}4\text{-}5, \qquad P_4: \ 2\text{-}6.$$

For notational convenience, let the event {e operates} be denoted by e and {e fails} by $\bar{e}$. Then $\Pr(E_1) = \Pr(15) = p_1 p_5$ is simple to compute, whereas the terms

$$\Pr(\bar{E}_1 E_2) = \Pr([\bar{1} \cup \bar{5}]136) = \Pr(\bar{5}136) = p_1 p_3 q_5 p_6$$

$$\Pr(\bar{E}_1 \bar{E}_2 E_3) = \Pr([\bar{1} \cup \bar{5}][\bar{1} \cup \bar{3} \cup \bar{6}]245)$$
$$= \Pr(\bar{1}[\bar{1} \cup \bar{3} \cup \bar{6}]245) = \Pr(\bar{1}245) = q_1 p_2 p_4 p_5$$

require liberal use of the laws of Boolean algebra. In processing the last term

$$\Pr(\bar{E}_1\bar{E}_2\bar{E}_3 E_4) = \Pr([\bar{1}\cup\bar{5}][\bar{1}\cup\bar{3}\cup\bar{6}][\bar{2}\cup\bar{4}\cup\bar{5}]26)$$
$$= \Pr([\bar{1}\cup\bar{5}][\bar{1}\cup\bar{3}][\bar{4}\cup\bar{5}]26)$$
$$= \Pr([\bar{1}\cup\bar{3}\bar{5}][\bar{4}\cup\bar{5}]26) = \Pr([\bar{1}\bar{4}\cup\bar{1}\bar{5}\cup\bar{3}\bar{5}]26),$$

the events $\bar{1}\bar{4}$, $\bar{1}\bar{5}$, and $\bar{3}\bar{5}$ which appear are not disjoint. We can continue by reapplying the disjoint-products formula to this new union of terms, yielding

$$\Pr(\bar{E}_1\bar{E}_2\bar{E}_3 E_4) = \Pr([\bar{1}\bar{4}\cup\bar{1}\bar{5}\cup\bar{3}\bar{5}]26) = \Pr([\bar{1}\bar{4}\cup 4\bar{1}\bar{5}\cup 1\bar{3}\bar{5}]26)$$
$$= \Pr(\bar{1}\bar{4}26 \cup 4\bar{1}\bar{5}26 \cup 1\bar{3}\bar{5}26)$$
$$= q_1 p_2 q_4 p_6 + q_1 p_2 p_4 q_5 p_6 + p_1 p_2 q_3 q_5 p_6.$$

The final expression for two-terminal reliability is then

$$R_{st}(G) = p_1 p_5 + p_1 p_3 q_5 p_6 + q_1 p_2 p_4 p_5 + q_1 p_2 q_4 p_6 + q_1 p_2 p_4 q_5 p_6 + p_1 p_2 q_3 q_5 p_6.$$

If, instead, the paths were presented in the order P_1, P_2, P_4, P_3, the calculations would be less demanding and we would obtain the simpler expression

$$R_{st}(G) = p_1 p_5 + p_1 p_3 q_5 p_6 + q_1 p_2 p_6 + p_1 p_2 q_3 q_5 p_6 + q_1 p_2 p_4 p_5 q_6.$$

This illustrates the effect of the chosen ordering on the complexity of the calculations.

A number of methods (Locks 1982; 1987) have been proposed to carry out the disjoint-products method, varying in their specific details but following the overall strategy outlined above. Typically, the paths P_i are first ordered by nondecreasing length and then processed in turn to generate a number of terms disjoint with one another and those previously generated. In general, the number of generated terms can grow rapidly with k, the number of given paths. However, there are identifiable cases in which each new path generates exactly one new term, so that the final reliability expression also has k terms. Systems that admit an ordering of paths with this property are called *shellable*, and there are efficient methods to generate the reliability given this ordering (Ball and Provan 1988). Such an ordering can be found efficiently in a number of cases, as when the system forms a *matroid*, a *nondegenerate linear system* or a *threshold system*. In particular, the disjoint-products method can be carried out efficiently, in terms of k, for the all-terminal reliability problem in undirected networks (a matroid) and the source-to-all-terminal reliability problem in directed networks (a nondegenerate linear system). No such efficient method is known, however, for calculating the two-terminal reliability measure.

2.4 Factoring

The inclusion–exclusion and disjoint-products techniques discussed in the previous section are based on a given enumeration of the s-t paths. By contrast, the *factoring* method does not require knowledge of these paths but instead concentrates on the state of an individual edge. Application of the pivotal decomposition formula (2.4) then creates two subproblems of smaller size that must now be solved. If the decomposition were simply reapplied to each such subproblem, the approach would be no better than state-space enumeration. Crucial to this approach is the possibility that certain of the generated subproblems might be reduced in size using simple probabilistic rules. We now present a few basic rules of reduction.

Two edges $e = (i, k)$ and $f = (i, k)$ joining the same two nodes in a directed network G are called *parallel* edges. A *parallel reduction* replaces two parallel edges, having reliabilities p_e and p_f, by a single edge having reliability $1 - (1 - p_e)(1 - p_f) = p_e + p_f - p_e p_f$. Two edges $e = (i, j)$ and $f = (j, k)$ are called *series* edges if these are the only two edges incident with node j. If $j \neq s, t$ then a *series reduction* replaces the two series edges by a single edge having reliability $p_e p_f$. Fig. 2.3 illustrates these two reliability-preserving reductions, which are valid in view of the independence of edge failures. Also illustrated is a more general *two-neighbour reduction*, applicable when $j \neq s, t$. A network G is *two-terminal series parallel* if it can be reduced to a single edge (s, t) by repeatedly applying series and parallel reductions. In such a case, the

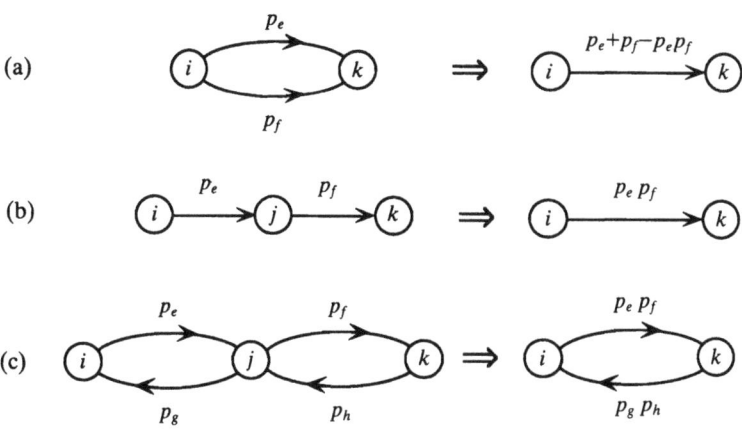

Fig. 2.3 Three probabilistic rules of reduction.

two-terminal reliability is just the reliability appearing on this final edge, and efficient algorithms exist for identifying and carrying out the appropriate reductions (Politof and Satyanarayana 1986). More generally, the application of series and parallel reductions to G will leave a network more complex than a single edge. At this point, an edge can be selected for conditioning and the pivotal decomposition formula can be applied, yielding two new subproblems. We would then like to apply series and parallel reductions to these subproblems for as long as possible, at which point pivotal decomposition can again be invoked. This alternating strategy of pivoting and applying reliability-preserving reductions constitutes the factoring algorithm.

For a directed network G, factoring on an edge e out of s, or into t, is especially helpful. Namely, the system G/e will now have a topological interpretation, being the network obtained from G by deleting edge e and merging its endpoints. While eqn (2.4) remains valid for any edge, unless the choice of edge for factoring is suitably restricted, G/e will not necessarily be equivalent to the network obtained from G by contracting the edge. This is clearly seen in the network of Fig. 2.1, since contraction of edge 3 would produce the spurious path 2-4 in Fig. 2.4(a). On the other hand, contraction of edge 1 produces the series–parallel network shown in Fig. 2.4(b) and its reliability is easily calculated as

$$R_{st}(G/e) = (p_2 p_5 + p_3 p_5 - p_2 p_3 p_5) + p_4 - (p_2 p_5 + p_3 p_5 - p_2 p_3 p_5) p_4.$$

Also, $G - e$ is accurately represented by the network of Fig. 2.1 with edge 1 removed. Since edges 3 and 4 are then irrelevant, they can be removed and $R_{st}(G - e) = p_2 p_5$. As a result of factoring on a single edge the two-terminal reliability of G is determined as

$$R_{st}(G) = p_1 R_{st}(G/e) + (1 - p_1) R_{st}(G - e)$$
$$= p_1 p_4 + p_2 p_5 + p_1 p_3 p_5 - p_1 p_2 p_3 p_5 - p_1 p_2 p_4 p_5 - p_1 p_3 p_4 p_5 + p_1 p_2 p_3 p_4 p_5.$$

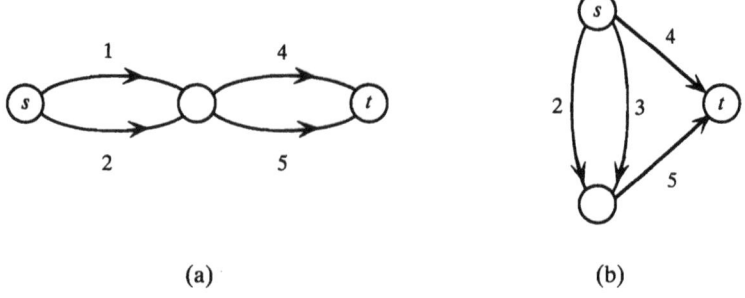

Fig. 2.4 Contraction of an edge in Fig. 2.1, using (a) $e = 3$ and (b) $e = 1$.

This example illustrates the potential advantages of decomposing a given network into smaller subnetworks each of which may admit certain probabilistic reductions, based on the network topology.

In the case of *undirected* networks, any edge can be used for factoring, but not all choices turn out to be equally effective. To formalize this idea, we view the factoring algorithm as a method for generating a binary computation tree T, whose root node corresponds to the original network G and in which the two descendants of any given node are the networks produced by factoring on some edge. Any such generated network will be simplified as much as possible using undirected series and parallel reductions, analogous to those given in Fig. 2.3(a) and (b). The leaf nodes of T (those having no descendants) consist of the completely resolvable subproblems; in general, these are networks with several connected components each of which contains either a single node or a single edge. One measure of the complexity of the factoring algorithm, relative to the particular edges used for conditioning at the various steps, is thus the number of leaves in T. Since the computation tree is binary, minimizing the number of leaves, $L(T)$, will also minimize the total number of subproblems generated, $2L(T) - 1$.

Satyanarayana and Chang (1983) showed that, apart from sign, the minimum number of leaves in any computation tree for G is precisely the coefficient $d(G)$ of $\Pr(G)$ appearing in eqn (2.5). This number, $D(G) = |d(G)|$, is called the *domination* of the undirected network G. To understand this connection, first notice that series and parallel reductions do not affect the domination of a network. Since a single edge $[s, t]$ has domination 1, it follows that $D(G) = 1$ for any series–parallel network. Moreover, the following result (Agrawal and Barlow 1984), derivable from eqn (2.6), shows that the domination is additive with respect to edge contraction and deletion in undirected networks.

Theorem 2.2. *If* $D(G) > 0$ *then* $D(G) = D(G/e) + D(G - e)$ *for any edge* e *of* G.

Because the domination is nonnegative the following corollary is immediate.

Corollary 2.3. $D(G) \leq D(G/e) + D(G - e)$ *for any edge* e *of* G.

Inductive use of this corollary shows that $D(G)$ is at most the sum of the dominations of the leaves of any computation tree T for G. However, the leaves have domination 0 or 1, with the value 1 occurring for networks containing a single edge $[s, t]$ and possibly some isolated nodes. Consequently, $L(T) \geq D(G)$ and, in view of Theorem 2.2, equality holds if every generated subproblem has positive domination. Satyanarayana and Chang (1983) discuss an edge-selection strategy that guarantees this positivity condition, so their edge-selection strategy is necessarily 'optimal' in terms of minimizing $L(T)$ over all possible computation trees T for G.

These ideas can be extended to the calculation of K-terminal reliability in undirected networks, in which appropriate series and parallel reductions are used to simplify networks before factoring on an edge. Alternative sets of reduction rules have also been studied, together with their associated optimal edge-selection strategies (Wood 1986). Agrawal and Barlow (1984) discuss, in further detail, the relation between factoring methods and domination, phrased in the more general realm of system reliability. Computer algorithms that carefully implement the factoring algorithm have been developed by several investigators (Page and Perry 1988, 1989; Resende 1988), and computational evidence indicates that the factoring algorithm is one of the most effective solution strategies for calculating the reliability of undirected networks.

2.5 Computational complexity

The previous sections have described three broad approaches for calculating network reliability: inclusion–exclusion, disjoint products, and factoring. Despite the considerable effort spent in devising various algorithms to implement these approaches, all known algorithms display a running time that grows exponentially with problem size. In this section we briefly indicate why network reliability problems are intrinsically difficult, in fact among the most challenging of all computational problems.

Ideally, one would like an algorithm whose running time grows polynomially with the 'size' of the input (for our purposes the number of nodes and edges in the network). Since not all input problems of the same size require the same computational effort, it is conventional to measure the time complexity of an algorithm by a function $f(z)$ that indicates its maximum running time on a problem instance of size z. A *decision problem* is one that asks whether a certain property holds or does not for the input instance: e.g., whether a given network contains a *Hamiltonian* cycle (which visits every node exactly once). If a decision problem can be decided by an algorithm with complexity function $f(z)$ bounded by a polynomial in z, then the problem belongs to the class P. A presumably broader class NP consists of those decision problems with the property that a purported solution can be *verified*, though not necessarily found, in polynomial time. The NP-complete problems represent the most difficult problems in NP; any polynomial-time algorithm for solving one NP-complete problem would enable all problems in the class NP to be solved in polynomial time. As yet, however, no one has discovered a polynomial-time algorithm for any NP-complete problem, although the list of such problems is large and the search has been intensive.

Reliability problems are more akin to counting problems than to decision problems or their optimization counterparts. A more appropriate problem class is that denoted by #P, consisting of all counting problems associated with decision problems in NP. For example, counting the number of Hamiltonian cycles in a network falls into the class #P. As a matter of fact, this specific problem is as hard as any problem in #P. It belongs to the problem class called #P-complete, characterized by the fact that a polynomial-time solution algorithm for such a problem could be efficiently transformed into a polynomial-time algorithm for all problems in #P. Another example of a #P-complete problem is counting the number of minimum cardinality s-t edge disconnecting sets in a network. Certainly #P-complete problems are at least as difficult as NP-complete problems, since knowing the number of solutions to a decision problem easily settles the existence question.

To provide a glimpse of the connection between reliability problems and such counting problems, consider the calculation of the two-terminal unreliability $U_{st}(G)$ given in eqn (2.3). Suppose that all m edges of G fail independently with probability $1-p$. Then

$$U_{st}(G) = \sum_{i=0}^{m} \mu_i (1-p)^i p^{m-i},$$

where μ_i is the number of s-t edge disconnecting sets in G of cardinality i, and thus

$$\sum_{i=0}^{m} \mu_i \left(\frac{1-p}{p}\right)^i = p^{-m} U_{st}(G). \tag{2.7}$$

Suppose we could calculate $U_{st}(G)$ efficiently for an arbitrary value p. By substituting $m+1$ distinct values of p, $0 < p < 1$, into eqn (2.7) we obtain a system of $m+1$ linear equations in $m+1$ unknowns μ_i with known right-hand sides. The coefficient matrix of this system is in fact a Vandermonde matrix with nonzero determinant, so that we could then solve for the μ_i values in polynomial time. In particular, we would be able to find μ_i for $i = \lambda_{st}(G)$, the size of the smallest s-t edge disconnecting set in G. Thus an efficient method for evaluating $U_{st}(G)$ would yield an efficient method for calculating the number of minimum cardinality s-t cutsets, known to be a #P-complete problem. We conclude that the evaluation of $R_{st}(G) = 1 - U_{st}(G)$, even when all edge reliabilities are identical, is an extremely difficult problem: in fact it too is #P-complete. It can be shown by similar means that calculation of the all-terminal reliability is #P-complete as well (Ball 1986).

Part of the inherent difficulty with approaches based on path enumeration, such as inclusion–exclusion and disjoint products, is that the number of paths can grow rapidly (in fact exponentially) with the problem size. It would be of interest then to know if there are methods for calculating reliability whose time complexity, while necessarily exponential in the problem size, grows only modestly with the number of paths or the number of cutsets. Rather surprisingly, there is an algorithm whose worst-case time complexity grows polynomially with the number of s-t cutsets, whereas there is unlikely to be one that grows polynomially with the number of s-t paths (Provan and Ball 1984). This cutset-based approach and extensions will be studied further in Chapter 6. To appreciate the reason for the apparent lack of such an algorithm based on s-t paths, consider a directed network $G = (N, E)$ with node set $N = N_1 \cup N_2 \cup \{s, t\}$, where N_1 and N_2 are disjoint. Network G contains edges (s, i) for $i \in N_1$, edges (j, t) for $j \in N_2$, and certain edges (i, j) for $i \in N_1$ and $j \in N_2$. The number of s-t paths in this network is the same as the number of edges joining nodes in N_1 to those in N_2. Since the number of s-t paths is then polynomial in the network size, a reliability algorithm polynomial in the number of s-t paths would be polynomial in the network size. The unlikelihood of this possibility follows from the #P-completeness (Provan and Ball 1984) of calculating $R_{st}(G)$ even for networks G with this special form (two-terminal directed bipartite networks).

2.6 Chapter notes

A good survey of various approaches for calculating the reliability of probabilistic networks is presented in Chapter 2 of Colbourn (1987a). Politof and Satyanarayana (1986) provide a survey of efficient methods that can be used for exact calculation of reliability in the special case of planar networks. In addition, the survey paper of Agrawal and Barlow (1984) emphasizes the key roles of factoring and domination in network reliability as well as in more general 'coherent' systems. Hwang et al. (1981) summarize a number of techniques that have been proposed for evaluating system reliability. A more detailed computational comparison of certain algorithms for the calculation of two-terminal reliability has been carried out by Yoo and Deo (1988).

While the state-space enumeration method of Section 2.1 is quite inefficient, it leads to a more competitive approach for the approximate (rather than the exact) calculation of reliability in networks. This topic is further developed in Chapter 8 of this book, where the approach is presented for general 'multistate' systems. The inclusion–exclusion method of Section 2.2 has been extended to the calculation of all-terminal reliability in directed networks by Satyanarayana and Hagstrom (1981a, b). A reduced inclusion–exclusion formula for K-terminal reliability problems has been developed by Satyanarayana (1982).

A number of algorithms have been proposed to implement the disjoint-products method. In addition to the algorithm of Locks (1987), the methods given by Hariri and Raghavendra (1987), Heidtmann (1989), and Wilson (1990) illustrate the computational efficacy of different ways of ordering and processing the given paths P_i. The factoring approach, discussed for undirected networks in Section 2.4, appears to have been first applied to directed networks by Nakazawa (1976). Reliability algorithms for directed networks that incorporate factoring, together with probabilistic reduction rules, have been implemented by Ball and Cameron (1986) and by Page and Perry (1989). Johnson (1984) and Wood (1986) discuss the application of the factoring approach to a variety of network reliability problems, in particular the K-terminal and all-terminal reliability problems for undirected networks.

A standard source for information on the computational complexity of algorithms is the book of Garey and Johnson (1979). More specific information on the complexity of network reliability problems and #P-complete problems can be found by consulting Rosenthal (1975) and Valiant (1979). An excellent summary of the appropriate complexity results is found in the papers of Ball (1980, 1986), as well as in Chapter 3 of Colbourn (1987a).

3
An algebraic formulation of network reliability problems

This chapter presents an algebraic approach for studying network reliability problems, concentrating, in particular, on the calculation of $R_{st}(G)$. Instead of working with numerical quantities (probabilities), the emphasis is on algebraic objects (polynomials) and their appropriate transformation. This algebraic perspective not only provides a unifying theoretical framework for studying network reliability problems, but also suggests a number of new algorithmic approaches. We will see, in particular, that several standard linear algebraic techniques can be readily adapted to solve network reliability problems. In this case, however, symbolic manipulations rather than numerical calculations will be carried out. The symbolic form of the resulting answer is especially useful in conducting sensitivity analyses, since the change in overall reliability with changes in component reliabilities can be directly ascertained from this form.

Suppose that $G = (N, E)$ is a directed network, with source node s and destination node t. The nodes of G are assumed to be perfect, whereas the $m = |E|$ edges are subject to failure. With each edge $k = 1, \ldots, m$ associate a variable x_k. It is desired to calculate the two-terminal *reliability polynomial* $R_{st}(\mathbf{x}) = R_{st}(x_1, \ldots, x_m)$, having the property that if the edge reliabilities p_k are substituted for the corresponding x_k then the resulting numerical value is precisely the two-terminal reliability. It will be seen in the next section that the reliability polynomial can be obtained by applying certain algebraic operations to the simple edge polynomials x_k. Notice that an undirected network H can be easily handled by associating two oppositely directed edges (i, j) and (j, i) with each undirected edge $[i, j]$ of H. Moreover, both directed edges receive the same edge variable x_k, indicating that the two edges both work or both fail together. More generally, it is possible to introduce certain degrees of dependence among the edge failures by judiciously labelling edges with more elaborate polynomials. For example, a common type of dependence occurs when nearby transmission links (e.g., those incident with the same node) are affected by the same weather or jamming equipment. This situation can be modelled by labelling each such edge with the composite polynomial $x_c x_k$, where x_c represents a common dependent failure variable and x_k represents an individual link failure variable.

3.1 Algebraic operations

In this section two relevant operations $\oplus$, $\otimes$ are defined on polynomials. To motivate their definition, first consider the case of two edges a and b in series, as depicted in Fig. 2.3(b). If the edges operate independently then their reliabilities p_a and p_b simply multiply. However, we need to allow for the possibility that the edges might also contain some common (dependent) information. Suppose, for example, that edge a is labelled $x_1 x_3 x_7$ and edge b is labelled $x_2 x_3 x_5$. Let A denote the event that components 1, 3, and 7 operate [$\Pr(A) = p_1 p_3 p_7$] and let B denote the event that components 2, 3, and 5 operate [$\Pr(B) = p_2 p_3 p_5$]. Thus AB is the event that 1, 2, 3, 5, and 7 all operate and $\Pr(AB) = p_1 p_2 p_3 p_5 p_7$, so that the single edge replacing a and b should be labelled $x_1 x_2 x_3 x_5 x_7$. More generally, let $w = x_{i_1} x_{i_2} \cdots x_{i_r}$ denote a *monomial term* and define the operation $\otimes$ on two such terms w^1, w^2 by:

$$w^1 \otimes w^2 = \prod \{\text{all } x_k \text{ in } w^1 \text{ or } w^2\}.$$

This operation can be extended to arbitrary polynomials by distributivity.

The second operation $\oplus$ is motivated by considering two edges a and b in parallel, as shown in Fig. 2.3(a). If the edges operated independently with probabilities p_a and p_b then their replacement edge would have reliability $p_a + p_b - p_a p_b$. This formula can likewise be extended to include the possibility of dependence. For example, suppose edge a is labelled $x_1 x_3 x_7$ and edge b is labelled $x_2 x_3 x_5$. Since $\Pr(A \cup B) = \Pr(A) + \Pr(B) - \Pr(AB)$ the single edge replacing a and b should receive the label $x_1 x_3 x_7 \oplus x_2 x_3 x_5 = x_1 x_3 x_7 + x_2 x_3 x_5 - x_1 x_2 x_3 x_5 x_7$. More generally, the algebraic sum of two polynomials f and g is defined by:

$$f \oplus g = f + g - (f \otimes g).$$

Operations related to $\oplus$ and $\otimes$ were apparently first suggested by Mine (1959) and by Kim et al. (1972). More recent development of these operations in the context of network reliability has been undertaken by Gondran and Minoux (1984) and Shier (1985).

Let $\mathscr{S}$ denote the set of all polynomials that can be formed from monomial terms by finite application of the operations $\oplus$ and $\otimes$. Then $(\mathscr{S}, \oplus, \otimes)$ forms a distributive lattice with smallest element 0 (the zero polynomial) and largest element 1 (the unit polynomial). Namely, for $f, g, h \in \mathscr{S}$

$f \oplus f = f$	$f \otimes f = f$	(3.1)
$f \oplus g = g \oplus f$	$f \otimes g = g \otimes f$	(3.2)
$f \oplus (g \oplus h) = (f \oplus g) \oplus h$	$f \otimes (g \otimes h) = (f \otimes g) \otimes h$	(3.3)

$$f \oplus (f \otimes g) = f \qquad\qquad f \otimes (f \oplus g) = f \qquad (3.4)$$
$$f \otimes (g \oplus h) = (f \otimes g) \oplus (f \otimes h) \qquad f \oplus (g \otimes h) = (f \oplus g) \otimes (f \oplus h) \qquad (3.5)$$
$$f \oplus 0 = f \qquad\qquad f \otimes 1 = f. \qquad (3.6)$$

An ordering $\leq$ on polynomials $f, g \in \mathcal{S}$ can be defined using the operation $\oplus$:

$$f \leq g \Leftrightarrow f \oplus g = g. \qquad (3.7)$$

This lattice ordering of polynomials is consistent with the normal ordering of real numbers. For instance, if $f = x_1$, $g = x_1 + x_2 - x_1 x_2$, then $f \oplus g = x_1 + (x_1 + x_2 - x_1 x_2) - (x_1 + x_1 x_2 - x_1 x_2) = g$. However, for any $0 \leq x_1, x_2 \leq 1$ we have $g = x_1 + x_2 - x_1 x_2 = x_1 + x_2(1 - x_1) \geq x_1 = f$, consistent with (3.7).

Using properties (3.1)–(3.6) it is then straightforward to show that for all $f, g, h \in \mathcal{S}$

$$0 \leq f \leq 1 \qquad (3.8)$$
$$f \otimes g \leq f \leq f \oplus g \qquad (3.9)$$
$$f \leq g \Rightarrow f \oplus h \leq g \oplus h \qquad (3.10)$$
$$f \leq g \Rightarrow f \otimes h \leq g \otimes h. \qquad (3.11)$$

To establish the connection between this algebraic structure and network reliability, we define $\mathcal{P}_{st}$ to consist of all simple s-t paths P in G. Also, define the *path value* $v(P)$ by the 'product' of the edge variables along the path P:

$$v(P) = \otimes \prod \{x_k : k \in P\}.$$

Then the reliability polynomial is expressible as

$$R_{st}(\mathbf{x}) = \oplus \sum_{P \in \mathcal{P}_{st}} v(P). \qquad (3.12)$$

In other words, the reliability polynomial is the 'sum' of path values $v(P)$ taken over all simple paths from s to t. This is simply a compact way of expressing the inclusion–exclusion formula for two-terminal reliability relative to the set of paths $\mathcal{P}_{st}$. Consider, for example, the bridge network of Fig. 2.2, with edges labelled using variables $x_1, \ldots, x_6$. Since there are four simple paths extending from node s to node t, application of (3.12) produces

$$R_{st}(\mathbf{x}) = x_1 x_5 \oplus x_2 x_6 \oplus x_1 x_3 x_6 \oplus x_2 x_4 x_5.$$

When this formula is expanded using the definitions of $\oplus$ and $\otimes$, the reliability polynomial becomes

$$R_{st}(\mathbf{x}) = x_1x_5 + x_2x_6 + x_1x_3x_6 + x_2x_4x_5 - x_1x_2x_5x_6 - x_1x_3x_5x_6 -$$
$$x_1x_2x_4x_5 - x_1x_2x_3x_6 - x_2x_4x_5x_6 + x_1x_2x_3x_5x_6 + x_1x_2x_4x_5x_6.$$

Substitution of the edge reliabilities $p_1, \ldots, p_6$ into this latter expression gives the two-terminal reliability of the system for any $0 \leq p_k \leq 1$. In particular, when all edges have the common edge reliability p, then the reliability polynomial becomes $2p^2 + 2p^3 - 5p^4 + 2p^5$.

3.2 Linear equations

The motivation for defining the operations $\oplus$ and $\otimes$, so that parallel edges 'add' and series edges 'multiply', can in fact be turned into a method for computing $R_{st}(\mathbf{x})$. The idea is to successively eliminate nodes from the network until only s and t remain. At the step in which node j is being eliminated, we simply apply the rule for series edges simultaneously to all edges (i, j) incident to j and all edges (j, k) incident from j. As a result of carrying out this step, certain other parallel and series reductions may then be possible. We illustrate this procedure on the example shown in Fig. 3.1(a), with edges labelled $x_1, \ldots, x_5$. The results of successively eliminating nodes a and b are summarized in the remainder of the figure. A final parallel reduction applied to Fig. 3.1(d)

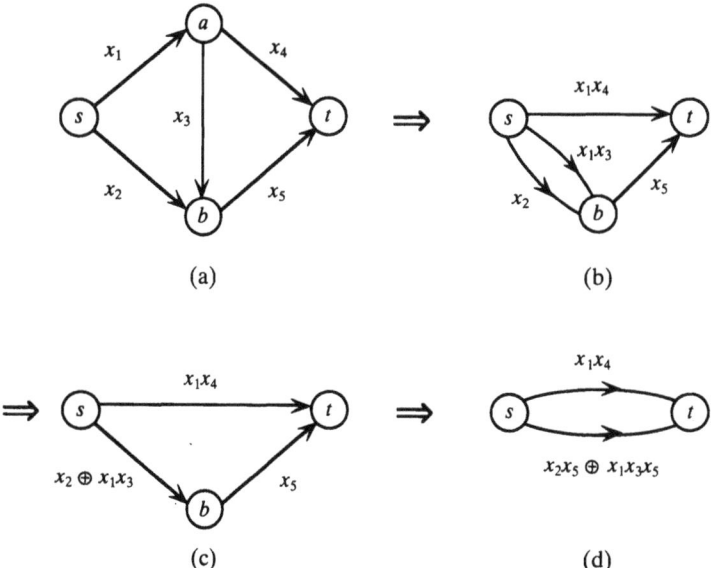

Fig. 3.1 Successive elimination of variables in a network.

produces the label $x_1 x_4 \oplus (x_2 x_5 \oplus x_1 x_3 x_5)$ on edge (s, t); when expanded using the definitions of $\oplus$ and $\otimes$ this yields the two-terminal reliability polynomial. It is significant that this elimination method does not require the simple s-t paths to be first enumerated. Such paths are automatically generated during the course of the computation itself.

It is also possible to associate a system of equations with a given network G. A quantity z_j is associated with each node j of G, and z_j can be viewed as a linear adder of inputs transmitted from other nodes adjacent to j. The label $x_k = x_{ij}$ associated with edge $k = (i, j)$ is considered to be an amplification factor that modifies the input from node i. Specifically, $z_s = 1$ and for $j \neq s$

$$z_j = \bigoplus_{(i,j) \in E} x_{ij} \otimes z_i. \qquad (3.13)$$

For example, the 'linear' equations associated with Fig. 3.1(a) are

$$z_s = 1$$
$$z_a = (x_1 \otimes z_s)$$
$$z_b = (x_2 \otimes z_s) \oplus (x_3 \otimes z_a)$$
$$z_t = (x_4 \otimes z_a) \oplus (x_5 \otimes z_b).$$

The appropriate interpretation of z_j is as the reliability polynomial $R_{sj}(\mathbf{x})$ relative to nodes s and j. Each $R_{sj}(\mathbf{x})$ depends on the s-j paths via (3.12), and the s-j paths for different nodes j are certainly related to one another. Consequently, there is an interdependence among the associated reliability polynomials, and this is captured by eqn (3.13). A more convincing justification for this relation will be given later. For now, it is important to point out how certain equations can be associated with a given network. Moreover, if z_a is now eliminated from the above equations by simple substitution we obtain the system

$$z_s = 1$$
$$z_b = (x_2 \otimes z_s) \oplus (x_1 x_3 \otimes z_s) = (x_2 \oplus x_1 x_3) \otimes z_s$$
$$z_t = (x_1 x_4 \otimes z_s) \oplus (x_5 \otimes z_b),$$

which corresponds exactly to the network in Fig. 3.1(c). Thus elimination of a node from the network has been manifested by elimination of the corresponding variable from the network equations.

In order to develop more formally this relation between linear systems and the reliability problem, it will be useful to define matrices relative to the algebraic structure $(\mathcal{S}, \oplus, \otimes)$. Let $n = |N|$ be the number of nodes in $G = (N, E)$ and let $\mathcal{M}_n$ denote the set of all $n \times n$ matrices with entries in $\mathcal{S}$. The usual notions of matrix addition $A \oplus B$ and multiplication $A \otimes B$ can then be defined over $\mathcal{M}_n$ relative to the operations $\oplus$ and $\otimes$. Properties (3.1)–(3.6) ensure that for matrices $A, B, C \in \mathcal{M}_n$:

$$A \oplus A = A$$
$$A \oplus B = B \oplus A$$
$$A \oplus (B \oplus C) = (A \oplus B) \oplus C \qquad A \otimes (B \otimes C) = (A \otimes B) \otimes C$$
$$A \otimes (B \oplus C) = (A \otimes B) \oplus (A \otimes C) \quad (B \oplus C) \otimes A = (B \otimes A) \oplus (C \otimes A)$$
$$A \oplus 0 = A \qquad A \otimes 1 = 1 \otimes A = A$$
$$A \otimes 0 = 0 \otimes A = 0$$

where 0 is the $n \times n$ matrix of all zero polynomials and 1 is the $n \times n$ diagonal matrix with unit polynomials along the diagonal and zero polynomials elsewhere. The order relation (3.7) can be extended as well to matrices using $A \leq B \Leftrightarrow A \oplus B = B$. The following analogues of (3.8)–(3.11) then hold for matrices in $\mathcal{M}_n$:

$$0 \leq A \leq A \oplus B$$
$$A \leq B \Rightarrow A \oplus C \leq B \oplus C$$
$$A \leq B \Rightarrow A \otimes C \leq B \otimes C, \quad C \otimes A \leq C \otimes B.$$

The powers A^j of matrix A are defined by successive matrix multiplications with respect to $\otimes$, with $A^0 = 1$. If, in addition, we define

$$A^* = \bigoplus_{j=0}^{\infty} A^j,$$

then the following three facts are easily established.

Lemma 3.1. $(A^*)^r = A^*$ for $r \geq 1$.

Lemma 3.2. $A \leq B \Rightarrow A^* \leq B^*$.

Lemma 3.3. $(A \oplus B)^* = A^*(B \otimes A^*)^*$.

Suppose, now, that $G = (N, E)$ is a given network with each edge $(i, j) \in E$ having the label x_{ij}. The *labelled adjacency matrix* $A = (a_{ij}) \in \mathcal{M}_n$ is defined by $a_{ij} = x_{ij}$ if $(i, j) \in E$ and $a_{ij} = 0$ otherwise. The fundamental observation is that the matrix $A^* = (a_{ij}^*)$ is precisely the matrix of reliability polynomials relative to all possible pairs of terminals.

Theorem 3.4. *If A is the labelled adjacency matrix for G then $A^* = (a_{ij}^*)$ satisfies*

$$a_{ij}^* = R_{ij}(\mathbf{x}).$$

Proof. If $i=j$ then $a_{ij}^* = 1 = R_{ii}(\mathbf{x})$, so suppose that $i \neq j$. Then the (i, j) element of A^r is given by

$$a_{ij}^r = \bigoplus \sum_{i_1,\ldots,i_{r-1}} a_{i,i_1} \otimes a_{i_1,i_2} \otimes \cdots \otimes a_{i_{r-1},j}$$

$$= \bigoplus \sum_{Q \in \mathcal{Q}_{ij}^r} v(Q),$$

where $\mathcal{Q}_{ij}^r$ indicates the set of all i-j paths in G of length r. As a result,

$$a_{ij}^* = \bigoplus \sum_{r=1}^{\infty} a_{ij}^r = \bigoplus \sum_{Q \in \mathcal{Q}_{ij}} v(Q),$$

where $\mathcal{Q}_{ij}$ indicates the set of all i-j paths in G. By eqn (3.12) $R_{ij}(\mathbf{x})$ is the corresponding sum taken over just the simple i-j paths. If, however, an i-j path Q is not simple then it can be decomposed into a simple i-j path P together with a number of cycles $C_1, \ldots, C_l$. Thus $v(Q) = v(P) \otimes v(C_1) \otimes \cdots \otimes v(C_l) \leq v(P)$ using the properties of $\leq$. Consequently, $v(Q) \oplus v(P) = v(P)$ and the inclusion of nonsimple paths in the expression for $R_{ij}(\mathbf{x})$ gives the same result as using only the simple paths, whence $a_{ij}^* = R_{ij}(\mathbf{x})$. □

From the proof of Theorem 3.4, it is seen that only the simple paths need to be included in the expression for A^*. Since any simple i-j path $(i \neq j)$ in a network with n nodes contains at most $n-1$ edges, it follows that the expression for A^* can be taken as the finite sum:

$$A^* = \bigoplus \sum_{j=0}^{n-1} A^j. \tag{3.14}$$

This explicit form for A^* establishes a connection between determining $R_{ij}(\mathbf{x})$ and solving the system of equations (3.13), which are now rewritten in a slightly more general form as

$$z = zA \oplus w. \tag{3.15}$$

Here A is the labelled adjacency matrix for G, z and w are row vectors over $\mathcal{S}$, and juxtaposition is used to indicate $\otimes$. The various properties of addition and multiplication extend in a natural way to vectors over $\mathcal{S}$. First note that $z^0 = wA^*$ solves (3.15), since $z^0 A \oplus w = (wA^*)A \oplus w = w(A^*A \oplus 1) = wA^*$. Conversely, suppose that z solves (3.15), so that $z = zA \oplus w = (zA \oplus w)A \oplus w = zA^2 \oplus w(1 \oplus A) = \cdots = zA^n \oplus w(1 \oplus A \oplus \cdots \oplus A^{n-1}) = zA^n \oplus wA^*$. Now if G is an acyclic network all paths have length at most $n-1$, whence $A^n = 0$ and $z = wA^*$ is the unique solution to (3.15). More generally, if G contains cycles, the system will have multiple solutions, but any such solution z satisfies $z = zA^n \oplus$

$wA^* \geqslant wA^*$, and so $z^0 = wA^*$ is the *minimal solution* of (3.15). The following result summarizes these findings.

Theorem 3.5. *Suppose A is the labelled adjacency matrix for G. Then $z^0 = wA^*$ is the minimal solution to $z = zA \oplus w$. If G is acyclic then z^0 is the unique solution to the system.*

As a result of Theorem 3.5, the two-terminal reliability problem can be considered as a linear algebra problem: solving a system of equations linear in the operations $\oplus$ and $\otimes$. Of course, the variables of the system are now polynomials in the edge labels. It is significant that the simple paths of G need not be enumerated in advance (as in other reliability methods), since their contribution to the reliability polynomial will be automatically accounted for in solving the system (3.15). Also, the network equations capture the interdependence that exists among the reliability polynomials $R_{sj}(\mathbf{x})$ for different nodes j. This interdependence can be exploited by various linear algebraic methods for calculating network reliability, as developed in the following sections.

While our concern is primarily with network reliability problems, it will be useful to point out the connection between such problems and other network path problems. For example, the classical shortest-path problem, in which an s-t path of minimum total length is sought, can be formulated in terms of two algebraic operations $\oplus$ and $\otimes$ applied to real numbers. Specifically, if $\oplus$ denotes the minimum operation and $\otimes$ denotes ordinary addition then the right-hand side of eqn (3.12) simply provides the length of the shortest s-t path: the minimum over all simple s-t paths P of the path value $v(P)$, which is now additive in the given edge lengths. This formulation remains valid for arbitrary real 'lengths' assigned to the edges of the network so long as every directed cycle in the network has nonnegative total length. Several other network problems (maximum capacity paths, minimax control paths, and kth shortest paths) can be viewed as algebraic path problems governed by operations $\oplus$ and $\otimes$ satisfying certain properties, similar to (3.1)–(3.6). An excellent account of this general viewpoint, and the resulting connection to solving linear systems, is given by Carré (1979).

3.3 Iterative methods

Our primary concern in this chapter has been the calculation of the two-terminal reliability polynomial $R_{st}(\mathbf{x})$. The appropriate set of network equations is then $z = zA \oplus e_s$, where e_s indicates the sth unit row vector with all components 0 except for the sth which equals 1. In view of Theorem 3.5, this system has the minimal solution $e_s A^*$, which is simply row s of the matrix A^*. Our initial interest in finding one specific component $R_{st}(\mathbf{x})$ of $e_s A^*$ will then be broadened to finding the entire row: the reliability polynomials $R_{sj}(\mathbf{x})$ for all nodes j. The solution methods developed here will in fact produce these reliability polynomials simultaneously.

The form of the equations $z = zA \oplus e_s$ suggests an *iterative* solution scheme for calculating the reliability polynomials $R_{sj}(\mathbf{x})$. Namely, the current estimate $z^{(r)}$ at iteration r is substituted into the right-hand side of the system, producing the estimate $z^{(r+1)}$ at the next iteration. This process is continued until an appropriate termination condition is met. Such a method represents an extension of the classical Jacobi method for solving ordinary linear systems. This *generalized Jacobi method* assumes an initial estimate $z^{(0)}$ and yields the sequence of estimates $\{z^{(r+1)}: r \geq 0\}$ according to

$$z^{(r+1)} = z^{(r)}A \oplus e_s. \tag{3.16}$$

Notice that if, for some r, $z^{(r-1)} \leq z^{(r)}$ then $z^{(r)} = z^{(r-1)}A \oplus e_s \leq z^{(r)}A \oplus e_s = z^{(r+1)}$. As a result the sequence $\{z^{(r)}: r \geq 0\}$ generated by the Jacobi method is nondecreasing if and only if $z^{(0)} \leq z^{(1)}$.

Repeated application of (3.16) yields an explicit expression for the iterates of this method: $z^{(1)} = z^{(0)}A \oplus e_s$, $z^{(2)} = z^{(1)}A \oplus e_s = (z^{(0)}A \oplus e_s)A \oplus e_s = z^{(0)}A^2 \oplus e_s(1 \oplus A), \ldots, z^{(r)} = z^{(0)}A^r \oplus e_s(1 \oplus A \oplus \cdots \oplus A^{r-1})$. By (3.14) it follows that $z^{(r)} = z^{(0)}A^r \oplus e_s A^* \geq e_s A^*$ for $r \geq n$. On the other hand, if $z^{(0)} \leq e_s A^*$ then $z^{(r)} = z^{(0)}A^r \oplus e_s A^* \leq e_s A^* A^r \oplus e_s A^* = e_s(A^*A^r \oplus A^*) = e_s A^*$. The following finite convergence result has thus been demonstrated.

Theorem 3.6. *The successive iterates of the generalized Jacobi method converge to $e_s A^*$ whenever $z^{(0)} \leq e_s A^*$. Convergence is assured in at most n iterations.*

While this theorem guarantees an upper bound of n on the required number of iterations, convergence can be detected whenever two successive iterates agree $z^{(r-1)} = z^{(r)}$. For, in that case, $z^{(r+1)} = z^{(r)}A \oplus e_s = z^{(r-1)}A \oplus e_s = z^{(r)}$, and so forth, giving $z^{(r-1)} = z^{(r)} = z^{(r+1)} = \cdots = z^{(n)} = e_s A^*$. Two particularly natural choices for $z^{(0)}$ are

(a) $z^{(0)} = e_s$,
(b) $z^{(0)} = e_s(1 \oplus A)$.

For both such initial vectors, the hypothesis of Theorem 3.6 is satisfied and convergence to $e_s A^*$ is assured. Indeed, since $A^* \geq 1$, $z^{(0)} = e_s \leq e_s A^*$, whereas $A^* \geq 1 \oplus A$ shows that $z^{(0)} = e_s(1 \oplus A) \leq e_s A^*$. In addition, it is straightforward to verify that $z^{(0)} \leq z^{(1)}$ holds for both choices, and thus the sequence of iterates produced in either case will be nondecreasing: $z^{(0)} \leq z^{(1)} \leq \cdots \leq z^{(n)} = e_s A^*$.

In practice, it can be computationally expensive to continue the Jacobi iterations until convergence occurs. An alternative is to terminate the process after several iterations, thus yielding a sequence of approximations to the reliability polynomials $R_{sj}(\mathbf{x})$. For the choices (a) and (b) above, the sequence of approximations is nondecreasing and the approximations $z^{(r)}$ provide better and better lower bounds on the actual network reliability. Notice that, regardless of the specific values for the edge reliabilities p_k, the approximation $z^{(r)}$ provides a 'conservative' estimate, guaranteeing that the given system will be at least as reliable as the approximate value produced.

Another well-known iterative scheme for solving linear systems can be adapted to solve the algebraic system $z = zA \oplus e_s$. To describe this method, suppose that the adjacency matrix A has been decomposed into a strictly upper triangular matrix U and a strictly lower triangular matrix L: $A = U \oplus L$. (Without loss of generality, it can be assumed that there are no edges of the form (i, i) since such *self loops* will not contribute to the overall reliability.) The matrices $U, L \in \mathcal{M}_n$ can then be viewed as the adjacency matrices of acyclic networks G_U and G_L, respectively. Given some initial row vector $z^{(0)}$, the *generalized Gauss–Seidel method* calculates the sequence of estimates $\{z^{(r+1)}: r \geq 0\}$ according to

$$z^{(r+1)} = z^{(r+1)}U \oplus z^{(r)}L \oplus e_s. \tag{3.17}$$

This method in effect bases the calculation of $z_j^{(r+1)}$ on $z_1^{(r+1)}, \ldots, z_{j-1}^{(r+1)}$ together with $z_{j+1}^{(r)}, \ldots, z_n^{(r)}$. By contrast, the Jacobi method bases the calculation of $z_j^{(r+1)}$ exclusively on $z_1^{(r)}, \ldots, z_n^{(r)}$. Since the components of the Gauss–Seidel vector are computed sequentially and not simultaneously (as in the Jacobi method), the convergence characteristics of this method can be sensitive to the actual numbering assigned to the nodes.

Because U is the adjacency matrix of an acyclic network G_U, Theorem 3.5 shows that the system (3.17) is equivalent to the system

$$z^{(r+1)} = z^{(r)}LU^* \oplus e_s U^*. \tag{3.18}$$

Thus the generalized Gauss–Seidel method amounts to carrying out the generalized Jacobi method on a network having adjacency matrix LU^*. Consequently, the Gauss–Seidel sequence $\{z^{(r)}: r \geq 0\}$ is nondecreasing if and only if $z^{(0)} \leq z^{(1)}$. Moreover, convergence is guaranteed whenever two successive iterates agree $z^{(r-1)} = z^{(r)}$.

An explicit expression for $z^{(r)}$ can be readily obtained by iterating the relation (3.18), yielding $z^{(r)} = z^{(0)}(LU^*)^r \oplus e_s U^*[1 \oplus LU^* \oplus \cdots \oplus (LU^*)^{r-1}]$. Suppose, now, that $r \geq n$. By (3.14) and Lemma 3.3, $z^{(r)} = z^{(0)}(LU^*)^r \oplus e_s U^*(LU^*)^* \geq e_s U^*(LU^*)^* = e_s(U \oplus L)^* = e_s A^*$. On the other hand, whenever $z^{(0)} \leq e_s A^*$ then $z^{(r)} = z^{(0)}(LU^*)^r \oplus e_s A^* \leq$

$e_s A^*(LU^*)^r \oplus e_s A^* \leq e_s A^*(A^*A^*)^r \oplus e_s A^* = e_s A^*$, where we have used $L \leq A \leq A^*$ and $U \leq A$, together with Lemmas 3.1 and 3.2. Hence $z^{(r)} = e_s A^*$ holds for $r \geq n$, establishing the following convergence result.

Theorem 3.7. *The successive iterates of the generalized Gauss–Seidel method converge to $e_s A^*$ whenever $z^{(0)} \leq e_s A^*$. Convergence is assured in at most n iterations.*

Consider again the choices of $z^{(0)}$ given by (a) or (b), previously shown to satisfy $z^{(0)} \leq e_s A^*$. For choice (a), $z^{(1)} = e_s(LU^*) \oplus e_s U^* \geq e_s U^* \geq e_s = z^{(0)}$. For choice (b), $z^{(1)} = e_s(1 \oplus A)LU^* \oplus e_s U^* \geq e_s(LU^* \oplus U^*) \geq e_s(L \oplus 1 \oplus U) = e_s(1 \oplus A) = z^{(0)}$. Either selection thus ensures a nondecreasing sequence of Gauss–Seidel iterates as well as finite convergence.

On intuitive grounds it appears reasonable that the Gauss–Seidel method, which utilizes more current information than the Jacobi method, should converge more rapidly. This can be rigorously established for certain choices of the initial vector $z^{(0)}$.

Theorem 3.8. *Consider the sequence of Jacobi iterates $w^{(r+1)} = w^{(r)} A \oplus e_s$ and the sequence of Gauss–Seidel iterates $z^{(r+1)} = z^{(r+1)} U \oplus z^{(r)} L \oplus e_s$, produced using the same initial vector $w^{(0)} = z^{(0)}$. If $z^{(0)} \leq z^{(1)}$ then $z^{(r)} \geq w^{(r)}$ for all $r \geq 0$.*

Proof. The result is established by induction on r. It clearly holds for $r = 0$ since $w^{(0)} = z^{(0)}$. Suppose that it holds for the value $r \geq 0$. Then $w^{(r+1)} = w^{(r)} A \oplus e_s \leq z^{(r)} A \oplus e_s = z^{(r)} U \oplus z^{(r)} L \oplus e_s \leq z^{(r+1)} U \oplus z^{(r)} L \oplus e_s = z^{(r+1)}$, so the result holds for $r + 1$. □

Since both initial vectors (a) and (b) satisfy the hypothesis $z^{(0)} \leq z^{(1)}$, then, for such choices, the Gauss–Seidel method will be uniformly at least as good as the corresponding Jacobi scheme. On the other hand, not all choices of $z^{(0)}$ will produce this behavior. To illustrate this possibility, consider the network of Fig. 3.2, in which $s = 1$. Using the initial approximation $w^{(0)} = (1, x_1 x_2, 0, 0, x_1 x_2 x_3, 0)$ the Jacobi scheme produces estimates $w^{(1)} = (1, 0, x_1 x_2 x_3 x_4, x_1, x_1 x_2 x_3, x_1 x_2 x_3)$ and $w^{(2)} = (1, x_1 x_2, x_1 x_2 x_3 x_4, x_1, x_1 x_2 x_3, x_1 x_2 x_3) = e_1 A^*$, the vector of reliability polynomials relative to source node 1. The Gauss–Seidel

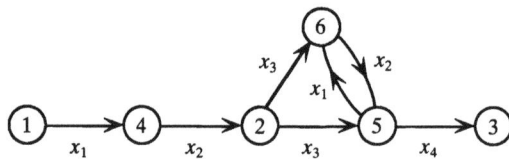

Fig. 3.2 An example for which Jacobi converges faster than Gauss–Seidel.

scheme with $z^{(0)} = w^{(0)}$ produces $z^{(1)} = (1, 0, x_1x_2x_3x_4, x_1, 0, 0)$, $z^{(2)} = (1, x_1x_2, 0, x_1, x_1x_2x_3, x_1x_2x_3)$, and the final estimate $z^{(3)} = (1, x_1x_2, x_1x_2x_3x_4, x_1, x_1x_2x_3, x_1x_2x_3) = w^{(2)}$. In this case, the Jacobi iterative method, which converges in two iterations, is superior to the Gauss–Seidel method, which requires three iterations.

3.4 Computational aspects

A specific iterative scheme is now discussed for solving the network equations $z = zA \oplus e_s$, where the emphasis is on identifying appropriate data structures and effective implementation strategies. In view of the common algebraic framework relating reliability problems to other network path problems, specifically the shortest-path problem, efficient solution methods developed for the latter problems can be adapted for use in reliability computations. After describing such a method for implementing the iterative approach, we develop certain simplifications that streamline the required computations. Numerical experience with the resulting approach is also reported.

Recall from (3.12) that the fundamental reliability computation involves processing the s-t paths in an efficient manner. The basic idea of the iterative method presented here is to keep, at each node i, a label $L(i)$, representing a reliability polynomial relative to a certain subset of paths from node s to node i. Since paths from node s to node i can be extended via the edge (i, j) to produce paths to node j, these labels can be updated in a forward manner. Namely, the information available at node i will be passed on to each adjacent node j through an appropriate updating formula.

To implement this idea, we maintain an ordered list $\mathscr{L}$ of nodes whose (polynomial) labels have been recently changed. Suppose that node i is removed from the list. Then the processing of node i will entail performing an (i, j)-update for each node j adjacent from i

$$L(j) := L(j) \oplus [L(i) \otimes x_k], \qquad (3.19)$$

where $k = (i, j)$ and x_k is the associated edge label. Any node j whose label is changed by the update (3.19) will be placed on $\mathscr{L}$, if not there already. This procedure of selecting and processing a node from the list $\mathscr{L}$ continues until $\mathscr{L}$ is empty.

Assuming an appropriate initialization of the labels $L(j)$, the update formula (3.19) ensures that $L(j)$ eventually incorporates all simple s-j paths plus possibly other nonsimple s-j paths. In view of the previous observation that no harm occurs if nonsimple paths are also included in the summation (3.12), it follows that when the procedure terminates each label $L(j)$ will equal $R_{sj}(\mathbf{x})$. Unlike the Jacobi and Gauss–Seidel methods described in the previous section, which process nodes in a fixed order, the present iterative method uses a dynamic, list-directed ordering on the nodes. That is, nodes are placed on the list only when their labels change and are processed as they are removed from the list. A more formal description of this iterative procedure, employing the initial approximation (b) of Section 3.3, is given below.

List-directed iterative algorithm. Given a network $G = (N, E)$, this algorithm produces $L(j) = R_{sj}(\mathbf{x})$ for all $j \in N$.

1. [Initialization]
   for $j \neq s$ do
      if $k = (s, j) \in E$ then $L(j) := x_k$
      else $L(j) := 0$;
   $L(s) := 1$; $\mathscr{L} := [j: (s, j) \in E]$.

2. [Iterative Step]
   while $\mathscr{L} \neq [\ ]$ do
      remove i from $\mathscr{L}$;
      for $k = (i, j) \in E$ do
         $T := L(j) \oplus [L(i) \otimes x_k]$;
         if $T \neq L(j)$ then
            $L(j) := T$; if $j \notin \mathscr{L}$ then enter j into $\mathscr{L}$.

In this implementation the label $L(j)$ corresponds to the current estimate of the reliability polynomial $R_{sj}(\mathbf{x})$. Each label is maintained as a fully expanded and simplified polynomial expression. To illustrate the mechanics of this approach, consider the bridge network of Fig. 2.2, in which the edge labels are $x_1, \ldots, x_6$. Initially $L(s) = 1$, $L(a) = x_1$, $L(b) = x_2$, $L(t) = 0$, and $\mathscr{L} = [a, b]$. Node a is then removed from $\mathscr{L}$ and the updates $L(b) = x_2 \oplus [L(a) \otimes x_3] = x_2 \oplus x_1 x_3 = x_2 + x_1 x_3 - x_1 x_2 x_3$, $L(t) = 0 \oplus [L(a) \otimes x_5] = x_1 x_5$ are performed. Now the list is $\mathscr{L} = [b, t]$ and node b is removed from $\mathscr{L}$, causing the updates $L(a) = x_1 \oplus [L(b) \otimes x_4] = x_1 + x_2 x_4 - x_1 x_2 x_4$, $L(t) = x_1 x_5 \oplus [L(b) \otimes x_6] = x_1 x_5 + x_2 x_6 + x_1 x_3 x_6 - x_1 x_2 x_3 x_6 - x_1 x_2 x_5 x_6 - x_1 x_3 x_5 x_6 + x_1 x_2 x_3 x_5 x_6$, $\mathscr{L} = [t, a]$. The processing continues until $\mathscr{L}$ becomes empty, at which point the polynomial label on each node represents the associated reliability polynomial.

In this example, the list $\mathscr{L}$ has been treated as a *queue*: nodes are sequentially entered at the 'bottom' of the list and sequentially removed from the 'top'. This technique for managing the list is called a first-in-first-out (FIFO) discipline, since it assures that nodes are processed in the same order in which they are entered. Alternatively, a last-in-first-out (LIFO) discipline would enter nodes at the top of the list and also remove nodes in turn from the top. This method of managing

Computational aspects 35

the list thus treats $\mathscr{L}$ as a *stack*, in which the most recently entered nodes are those first used for subsequent processing. The effect of these two ways of managing $\mathscr{L}$ will be examined later, particularly as it influences the quality of the approximations produced.

Certain useful properties of this iterative algorithm will now be established. Each *step* of the algorithm corresponds to removing a node from $\mathscr{L}$, and $L_k(j)$ denotes the polynomial label on node j at the beginning of step k. It will be convenient to let x denote the label on edge (i, j). Thus the (i, j)-update of node j during step k can be expressed as

$$L_{k+1}(j) := L_k(j) \oplus xL_k(i),$$

where the product $\otimes$ is indicated by juxtaposition. Iterating the above relation shows that if $v \leq w$ then $L_w(j) = L_v(j) \oplus Z \succcurlyeq L_v(j)$, which we record as

Property 3.9. *If $v \leq w$ then $L_v(j) \leq L_w(j)$.*

An important simplification to the calculations derives from the fact that only the 'new' information $N(i)$ added to the label of node i since i was last processed needs to be propagated to adjacent nodes.

Property 3.10. *Suppose that an (i, j)-update of node j occurs during step k, where the labels on i and j prior to the update are $L_k(i) = L_r(i) \oplus N(i)$ and $L_k(j)$, respectively. That is, an (i, j)-update to node j previously occurred at step $r < k$. Then at step $k+1$ the new label on node j will be $L_{k+1}(j) = L_k(j) \oplus xN(i)$.*

Proof. At step r, node j receives the label $L_{r+1}(j) = L_r(j) \oplus xL_r(i)$. Since $r < k$ Property 3.9 shows that $L_k(j) \oplus L_{r+1}(j) = L_k(j)$, so that

$$\begin{aligned}
L_{k+1}(j) &= L_k(j) \oplus xL_k(i) \\
&= [L_k(j) \oplus L_{r+1}(j)] \oplus x[L_r(i) \oplus N(i)] \\
&= [L_k(j) \oplus L_r(j) \oplus xL_r(i)] \oplus xL_r(i) \oplus xN(i) \\
&= [L_k(j) \oplus L_r(j) \oplus xL_r(i)] \oplus xN(i) \\
&= [L_k(j) \oplus L_{r+1}(j)] \oplus xN(i) \\
&= L_k(j) \oplus xN(i). \quad \square
\end{aligned}$$

Because the labels on each node are maintained as fully expanded polynomials (expressed using ordinary $+$ and $\times$), it is desirable to know when certain of the terms present in $L_k(i)$ do not affect the label $L_k(j)$ during an (i, j)-update. The following property provides such a condition.

Property 3.11. *Suppose an (i, j)-update is performed at step k where $L_k(i) = Y_1 + Y_2 + \cdots + Y_v$, $L_k(j) = Z_1 \oplus Z_2 \oplus \cdots \oplus Z_w$, and $Z_1 \succcurlyeq xY_1$.*

Then the updated label $L_{k+1}(j) = L_k(j) \oplus x[Y_2 + \cdots + Y_v]$.

Proof. Let $Y = Y_2 + \cdots + Y_v$ and $Z = Z_2 \oplus \cdots \oplus Z_w$. Since $Z_1 \geqslant xY_1$, it follows from eqn (3.4) that $Z_1 \otimes xY_1 = (Z_1 \oplus xY_1) \otimes xY_1 = xY_1$, whence

$$\begin{aligned}
L_{k+1}(j) &= L_k(j) \oplus xL_k(i) \\
&= [Z \oplus Z_1] \oplus x[Y_1 + Y] \\
&= Z \oplus [Z_1 \oplus (xY_1 + xY)] \\
&= Z \oplus [Z_1 + xY_1 + xY - (Z_1 \otimes xY_1) - (Z_1 \otimes xY)] \\
&= Z \oplus [Z_1 + xY_1 + xY - xY_1 - (Z_1 \otimes xY)] \\
&= Z \oplus [Z_1 + xY - (Z_1 \otimes xY)] \\
&= Z \oplus [Z_1 \oplus xY] \\
&= L_k(j) \oplus x[Y_2 + \cdots + Y_v]. \quad \square
\end{aligned}$$

Property 3.9 shows that the labels produced at any node by the iterative algorithm form a nondecreasing sequence of approximations to $R_{sj}(\mathbf{x})$. Properties 3.10 and 3.11 show that certain cancellations in the update step (3.19) can be predicted in advance, and thus unnecessary computations can be avoided. Empirical results from using an iterative algorithm that incorporates these refinements will now be presented.

The first test network, having 9 nodes and 19 edges, is taken from Satyanarayana and Prabhakar (1978) and is shown in Fig. 3.3. There are 35 s-t paths and 5 287 noncancelling terms in $R_{st}(\mathbf{x})$; as discussed in Section 2.2 each noncancelling term corresponds to an acyclic subnetwork of G. Despite its small size, this example represents one of the most complex directed networks whose exact reliability has been reported in the literature.

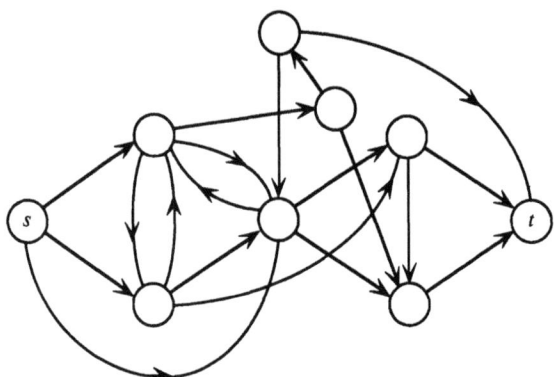

Fig. 3.3 Example network of Satyanarayana and Prabhakar.

The reliability polynomials $R_{sj}(\mathbf{x})$ for this network have been calculated using the iterative procedure described in this section. For ease of presentation, our attention will focus on the sequence of approximations to $R_{st}(G)$, evaluated at the common edge reliability p. Fig. 3.4 displays the sequence of approximations L_k produced using the FIFO discipline. As expected, the reliability curves L_k form a nondecreasing sequence that converges to the exact answer, requiring 9 iterations. (Each iteration corresponds to a change in the label on node t.) Notice that the curves for the fifth to ninth iterations closely overlap in the figure, thus providing excellent approximations to the exact reliability. The figure also indicates the cumulative CPU times (in seconds, IBM 3081) required to complete the work to the end of the specified iteration. Thus a total of 0.638 seconds were needed to obtain $R_{st}(G)$, whereas only 0.061 seconds were needed to obtain an approximation that is virtually indistinguishable over the entire range $0 \le p \le 1$.

Fig. 3.5 shows the approximations obtained when the LIFO discipline is used in the iterative algorithm. In this case, twelve iterations are required before convergence is obtained. (Several of the curves overlap so only 10 approximations are apparent in the figure.) Although the exact answer was obtained in 0.454 seconds, less than the comparable time for FIFO, the LIFO discipline did not give as useful a set of approximations compared to the FIFO approach.

Additional computational results are presented for five randomly

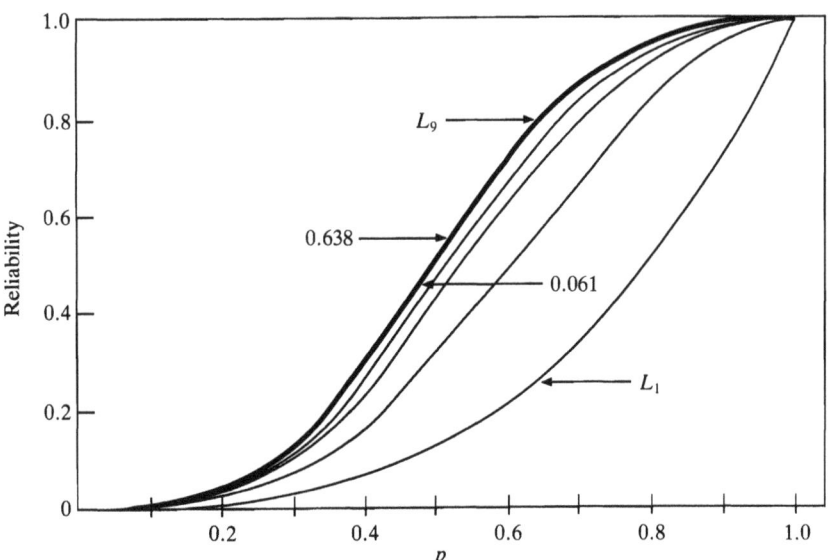

Fig. 3.4 Approximations to reliability using the FIFO discipline.

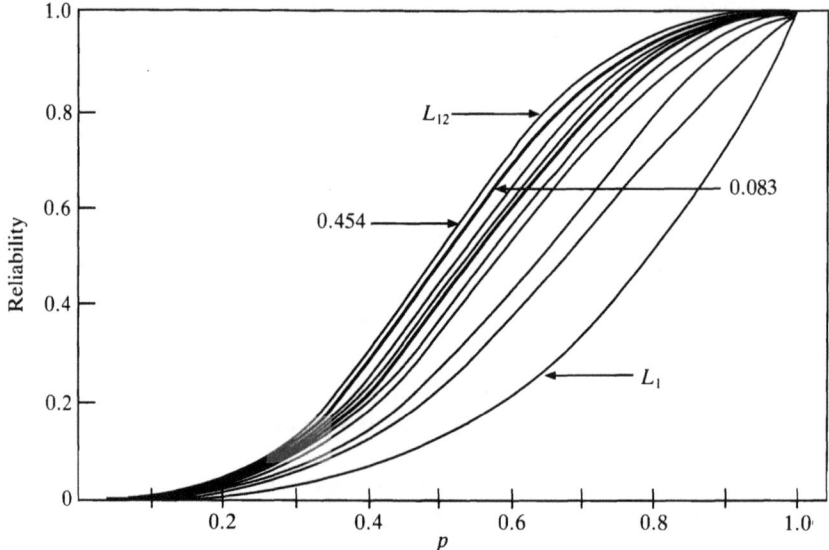

Fig. 3.5 Approximations to reliability using the LIFO discipline.

generated networks having 12 nodes and 30 edges. Certain characteristics of these networks, together with the number of iterations required for convergence, are shown in Table 3.1. In order to compare the quality of the approximations produced for these examples, we have tabulated the CPU time (in seconds) required to achieve a relative error of $\alpha\%$ or less (at $p = 0.5$) in Table 3.2. For these random examples, the FIFO and LIFO disciplines appear to be roughly comparable in terms of the time required to obtain the exact answer. However, the FIFO variant consistently gives close approximations with only a modest computational effort and it completely dominates the LIFO variant in this respect.

These and other empirical results demonstrate that the choice of list

Table 3.1 Iterative algorithms applied to five random networks (12 nodes, 30 edges)

Network	Number of s-t paths	Number of noncancelling terms	Number of iterations FIFO	LIFO
R1	14	1 263	9	7
R2	28	3 383	11	8
R3	41	7 583	8	10
R4	44	17 919	5	5
R5	34	42 687	10	8

Table 3.2 Comparison of the accuracy of FIFO and LIFO for random networks

Example	Discipline	\multicolumn{4}{c}{CPU (secs) for accuracy within $\alpha\%$}			
		0%	1%	5%	10%
R1	FIFO	0.049	0.001	0.001	0.001
	LIFO	0.054	0.054	0.004	0.001
R2	FIFO	0.380	0.076	0.006	0.003
	LIFO	0.362	0.362	0.362	0.362
R3	FIFO	1.21	0.172	0.172	0.004
	LIFO	1.11	0.630	0.162	0.071
R4	FIFO	3.06	0.195	0.004	0.004
	LIFO	2.85	2.85	0.089	0.005
R5	FIFO	5.07	0.062	0.002	0.001
	LIFO	5.11	1.58	1.58	0.002

structure (FIFO, LIFO) can have a significant effect on the relative efficiency of the procedures as well as on the quality of the approximations. Whereas the LIFO approach tends to find the exact reliability polynomial somewhat faster than the FIFO approach, the FIFO approach produces better approximations. This desirable feature of the FIFO approach can be explained as follows, assuming that the edge reliabilities are comparable in magnitude. Under a FIFO discipline, nodes are processed in order of increasing distance from s. Thus the first time node j receives a label, it is done so relative to a path having the minimum number of edges in an s-j path. In general, the FIFO approach ensures that the more probable paths (having fewer edges) are incorporated as soon as possible. Subsequently the smaller contributions of the less probable and longer paths are included. On the other hand, a LIFO discipline creates a depth-first rather than a breadth-first search of the network, and thus early approximations can be substantially improved by the incorporation of later (shorter) paths.

3.5 Chapter notes

This chapter has developed an algebraic structure, defined over a certain space of polynomials, that is appropriate for the two-terminal reliability problem. Use of this algebraic structure provides a compact representation of the reliability polynomial $R_{st}(\mathbf{x})$ in terms of two fundamental operations. A similar type of algebraic structure governs other types of path problems in networks: shortest paths, maximum capacity paths, kth best paths, and multicriteria paths, for example. Carré (1971, 1979), Gondran and Minoux (1984), and Zimmermann (1981) discuss diverse applications of algebraic path problems and develop a unifying algebraic structure. The use of linear equations to represent and solve such path problems is also emphasized in these references. A variety of linear algebraic techniques for solving algebraic path problems have been developed and compared by Carré (1971, 1979) and Shier (1973, 1976). Application of the algebraic structure appropriate for two-terminal reliability calculations and the development of specific iterative algorithms can be found in Shier (1985) and Shier and Whited (1987). Additional computational results employing the list-directed algebraic algorithm of Section 3.4 are reported in Shier and Whited (1987).

4

Bounds on two-terminal reliability

As indicated in Section 2.5, the exact calculation of network reliability is #P-complete, meaning that an efficient method is unlikely to exist for carrying out this computation. Consequently, it is appropriate to discuss method for approximating network reliability. In particular, this chapter will discuss a number of methods for bounding the two-terminal reliability of a network. It will be useful to classify such bounds as either *static* or *dynamic* bounds. The lower-bound polynomials developed in Sections 3.3–3.4 are examples of dynamic bounds because they provide a *sequence* of approximations to the exact reliability. Section 4.1 surveys a variety of lower and upper bounds on reliability, while Sections 4.2–4.3 concentrate on bounds that derive from algebraic considerations. These algebraic bounds are dynamic estimates that are guaranteed to converge monotonically to the exact two-terminal reliability polynomial in a finite number of iterations. Moreover, the algebraic approach can be implemented to produce matched lower and upper bounds at each iteration. Thus at each step one can compute an interval guaranteed to contain the correct reliability value, and the width of this interval will progressively decrease. As a result, it is possible to monitor the interval of uncertainty and terminate the iterative process once the current interval is sufficiently small or the allocated computing resources have been expended.

4.1 Bounding techniques

Throughout, it is supposed that $G = (N, E)$ is a directed network, with source node s and destination node t. Our objective here will be to obtain lower and upper bounds on the two-terminal reliability $R_{st}(G)$, the probability that there is a path of operative edges from s to t. Let $P_1, P_2, \ldots, P_k$ denote the simple s-t paths of G, and let E_i be the event that all edges in path P_i operate. Then the two-terminal reliability is given by eqn (2.2):

$$R_{st}(G) = \Pr(E_1 \cup E_2 \cup \cdots \cup E_k).$$

As in Section 2.2, this expression can be expanded using the principle of inclusion and exclusion to obtain an alternating sum of terms. In fact, it can be shown (Barlow and Proschan 1981; Ross 1985) that the cumulative sums (based on single events, at most two events, at most three events, ...) bound the exact value alternately from above and below. Specifically, if S_w denotes the sum of $\Pr(E_{i_1}E_{i_2}\cdots E_{i_w})$ over all $i_1 < i_2 < \cdots < i_w$ then

$$R_{st}(G) \le S_1, \quad R_{st}(G) \ge S_1 - S_2, \quad R_{st}(G) \le S_1 - S_2 + S_3, \ldots . \quad (4.1)$$

Such bounds are clearly dynamic, since we obtain a sequence of lower bounds and a sequence of upper bounds. Yet it is not necessarily the case that the lower-bound sequence is monotone nondecreasing and that the upper-bound sequence is monotone nonincreasing (Schwager 1984). These *inclusion–exclusion bounds,* also called the Bonferroni bounds, are clearly valid for any input edge reliabilities p_i. However, the bounds developed in (4.1) are only practically useful when the p_i are small, since then the omitted terms $\Pr(E_{i_1}E_{i_2}\cdots E_{i_w})$ will be small and the correction factors S_w will also tend to be small. Inclusion–exclusion bounds that are useful for larger values of the p_i (the more common situation) can be obtained by focusing, instead, on the s-t cutsets $C_1, C_2, \ldots, C_r$. If F_j denotes the event that all edges in cutset C_j fail, then the two-terminal unreliability of G is given by eqn (2.3):

$$U_{st}(G) = 1 - R_{st}(G) = \Pr(F_1 \cup F_2 \cup \cdots \cup F_r).$$

In a similar way, the two-terminal unreliability can be approximated in terms of T_h, the sum of $\Pr(F_{i_1}F_{i_2}\cdots F_{i_h})$ over all $i_1 < i_2 < \cdots < i_h$:

$$U_{st}(G) \le T_1, \quad U_{st}(G) \ge T_1 - T_2, \quad U_{st}(G) \le T_1 - T_2 + T_3, \ldots . \quad (4.2)$$

Nelson *et al.* (1970) discuss a computer implementation of this approach for bounding the two-terminal reliability of a network. An improved set of bounds, based on computing certain 'binomial moments,' can alternatively be obtained from a linear-programming formulation (Prékopa 1988; Prékopa 1990).

To illustrate the use of such bounds, consider the example network G in Fig. 4.1. For ease of presentation, suppose that all edges fail independently with the same failure probability $q = 1 - p$. The simple s-t

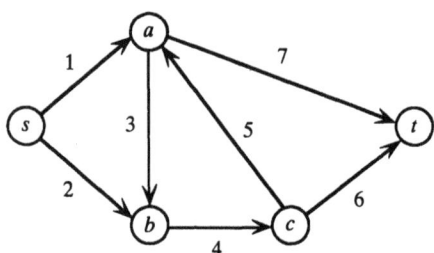

Fig. 4.1 An illustrative network.

paths of G are

P_1: 1-7, $\quad P_2$: 2-4-6, $\quad P_3$: 1-3-4-6, $\quad P_4$: 2-4-5-7

and the s-t cutsets of G are

C_1: 1-2, $\quad C_2$: 4-7, $\quad C_3$: 6-7, $\quad C_4$: 1-4, $\quad C_5$: 2-3-7,
$\quad C_6$: 1-5-6.

Then the inclusion–exclusion bounds based on paths are

$$IEP_{U1} = S_1 = p^2 + p^3 + 2p^4$$
$$IEP_{L1} = S_1 - S_2 = p^2 + p^3 + 2p^4 - 5p^5 - p^7$$
$$IEP_{U2} = S_1 - S_2 + S_3 = p^2 + p^3 + 2p^4 - 5p^5 + 2p^6 + p^7$$
$$IEP_{L2} = R_{st}(G) = p^2 + p^3 + 2p^4 - 5p^5 + 2p^6.$$

Table 4.1 shows the numerical values of these upper and lower bounds, together with the exact reliability value, for selected values of p. In this example, the bounds are seen to be quite satisfactory for small values of the common edge reliability p.

The bounds on unreliability given in (4.2) produce here the following cutset-based bounds on two-terminal reliability:

$$IEC_{L1} = 1 - T_1 = 1 - 4q^2 - 2q^3$$
$$IEC_{U1} = 1 - T_1 + T_2 = 1 - 4q^2 + q^3 + 9q^4 + 2q^5 + q^6$$
$$IEC_{L2} = 1 - T_1 + T_2 - T_3 = 1 - 4q^2 + q^3 + 7q^4 - 10q^5 - 3q^6 - 2q^7$$
$$IEC_{U2} = 1 - T_1 + T_2 - T_3 + T_4 = 1 - 4q^2 + q^3 + 7q^4 - 7q^5 + 4q^6 + 3q^7$$
$$IEC_{L3} = 1 - T_1 + T_2 - T_3 + T_4 - T_5 = 1 - 4q^2 + q^3 + 7q^4 - 7q^5 + 2q^6 - q^7$$
$$IEC_{U3} = R_{st}(G) = 1 - 4q^2 + q^3 + 7q^4 - 7q^5 + 2q^6.$$

Table 4.1 Inclusion–exclusion bounds based on paths

p	IEP_{L1}	$R_{st}(G)$	IEP_{U2}	IEP_{U1}
0.0	0.0	0.0	0.0	0.0
0.1	0.01115	0.01115	0.01115	0.01120
0.2	0.04959	0.04973	0.04974	0.05120
0.3	0.12083	0.12251	0.12273	0.13320
0.4	0.22236	0.23219	0.23383	0.27520
0.5	0.33594	0.37500	0.38281	0.50000
0.6	0.41841	0.53971	0.56771	0.83520

Table 4.2 shows the numerical values of these upper and lower bounds, together with the exact reliability value, for selected values of $q = 1 - p$. In this example, the cutset-based bounds are reasonably good for large values of p (small values of q). However, for small values of p, the calculated estimates are egregiously bad—indeed certain estimates are negative or greater than 1. Likewise, the path-based bounds can produce quite uninformative estimates if p is sufficiently large.

These inclusion–exclusion bounds become increasingly expensive to compute with each further iteration. Therefore we now explore an easily computed static bound that incorporates the structure of the paths and cutsets in a different manner. Again, suppose that E_i is the event in which all edges of path P_i operate. In general, these s-t paths share common edges, so the associated events E_i are statistically dependent. However, by making the optimistic assumption that they are independent, we will, in fact, obtain an upper bound on the operating probability of the system. Namely

$$R_{st}(G) = \Pr(E_1 \cup E_2 \cup \cdots \cup E_k)$$

$$= 1 - \Pr(\bar{E}_1 \bar{E}_2 \cdots \bar{E}_k) \leq 1 - \prod_{i=1}^{k} \Pr(\bar{E}_i) = 1 - \prod_{i=1}^{k} [1 - \Pr(E_i)].$$

Of course, each $\Pr(E_i)$ is easy to calculate since the edge failures are presumed independent. The validity of this argument was first established by Esary and Proschan (1963), and so we refer to the final quantity on the right above as the *Esary–Proschan upper bound* EP_U. In a similar way, consideration of the cutsets C_j and their associated events F_j leads to the *Esary–Proschan lower bound* EP_L on reliability:

$$R_{st}(G) = 1 - \Pr(F_1 \cup F_2 \cup \cdots \cup F_r)$$

$$= \Pr(\bar{F}_1 \bar{F}_2 \cdots \bar{F}_r) \geq \prod_{j=1}^{r} \Pr(\bar{F}_j) = \prod_{j=1}^{r} [1 - \Pr(F_j)].$$

Table 4.2 Inclusion–exclusion bounds based on cutsets

q	IEC_{L1}	IEC_{L2}	IEC_{L3}	$R_{st}(G)$	IEC_{U2}	IEC_{U1}
0.0	1.0	1.0	1.0	1.0	1.0	1.0
0.1	0.95800	0.96160	0.96163	0.96163	0.96163	0.96192
0.2	0.82400	0.85578	0.85708	0.85709	0.85725	0.86310
0.3	0.58600	0.69678	0.70793	0.70815	0.71026	0.74549
0.4	0.23200	0.48524	0.53807	0.53971	0.55282	0.67898
0.5	−0.25000	0.18750	0.36719	0.37500	0.42969	0.76563
0.6	−0.87200	−0.29036	0.20420	0.23219	0.40949	1.14458

Table 4.3 Static bounds

p	DP_L	EP_L	$R_{st}(G)$	EP_U	DC_U
0.0	0.0	0.0	0.0	0.0	0.0
0.1	0.01099	0.00010	0.01115	0.01119	0.03610
0.2	0.04768	0.00400	0.04973	0.05073	0.12960
0.3	0.11457	0.02920	0.12251	0.12886	0.26010
0.4	0.21376	0.10312	0.23219	0.25350	0.40960
0.5	0.34375	0.24225	0.37500	0.42322	0.56250
0.6	0.49824	0.43618	0.53971	0.61987	0.70560
0.7	0.66493	0.64922	0.70815	0.80651	0.82810
0.8	0.82432	0.83581	0.85709	0.93876	0.92160
0.9	0.94851	0.95868	0.96163	0.99391	0.98010
1.0	1.0	1.0	1.0	1.0	1.0

The upper bound is most useful when the p_i are small whereas the lower bound is appropriate when the p_i are relatively large (Barlow and Proschan 1981). Jensen and Bellmore (1969) discuss a computer implementation of this approach, including a method for generating the cutsets, and they also report empirical results from a number of test problems.

We illustrate the application of these static bounds to the example in Fig. 4.1, assuming for convenience that all edge reliabilities equal p. Considering the four s-t paths listed earlier gives $\Pr(E_1) = p^2$, $\Pr(E_2) = p^3$, $\Pr(E_3) = \Pr(E_4) = p^4$ and the upper bound $EP_U = 1 - (1 - p^2)(1 - p^3)(1 - p^4)^2$. Using the six s-t cutsets produces $\Pr(F_1) = \Pr(F_2) = \Pr(F_3) = \Pr(F_4) = q^2$, $\Pr(F_5) = \Pr(F_6) = q^3$ and the lower bound $EP_L = (1 - q^2)^4(1 - q^3)^2$. Numerical values for these bounds at various p are found in the appropriate columns of Table 4.3. Note that, unlike the inclusion-exclusion bounds, the Esary–Proschan bounds define genuine reliability polynomials. Indeed, the upper bound represents the reliability polynomial of a system that has the paths of G arranged in parallel (Fig. 4.2a), while the lower bound represents the reliability polynomial of a system that has the cutsets of G placed in series (Fig. 4.2b).

Other static bounds can be deduced from the s-t paths and cutsets of a network. The basic idea is that since the events E_i (or F_j) are not in general disjoint we can focus on a subset of these events that are disjoint and thereby obtain bounds on the overall reliability. Specifically, suppose that the simple s-t paths $P_1, P_2, \ldots, P_k$ are arranged so that the first w such paths are edge disjoint. Then

$$R_{st}(G) = \Pr(E_1 \cup E_2 \cup \cdots \cup E_k)$$

$$\geq \Pr(E_1 \cup E_2 \cup \cdots \cup E_w)$$
$$= 1 - \Pr(\bar{E}_1 \bar{E}_2 \cdots \bar{E}_w)$$
$$= 1 - \prod_{i=1}^{w} [1 - \Pr(E_i)],$$

where the last equality follows from the disjointness of the paths. In this formula for the *disjoint-paths lower bound* DP_L, it is clearly desirable to make the w-term product above as small as possible. Ideally, we could accomplish this by finding a large number of disjoint paths each having few edges. In practice, there can be trade-offs between these two criteria. Various heuristic procedures for selecting a 'good' collection of paths have been proposed by Brecht and Colbourn (1988). Their computational results indicate that these heuristics produce lower bounds that are competitive with the best of the existing path-based bounds.

By considering instead the s-t cutsets $C_1, C_2, \ldots, C_r$ of G and choosing the first h to be edge disjoint, the following *disjoint-cutsets upper bound* DC_U results:

$$R_{st}(G) = 1 - \Pr(F_1 \cup F_2 \cup \cdots \cup F_r)$$
$$= \Pr(\bar{F}_1 \bar{F}_2 \cdots \bar{F}_r)$$
$$\leq \Pr(\bar{F}_1 \bar{F}_2 \cdots \bar{F}_h)$$
$$= \prod_{j=1}^{h} [1 - \Pr(F_j)].$$

In order to make this upper bound as small as possible, it is desirable to select a large number of disjoint cutsets, each involving relatively few

(a)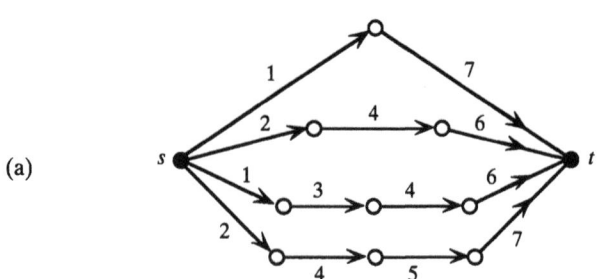

(b)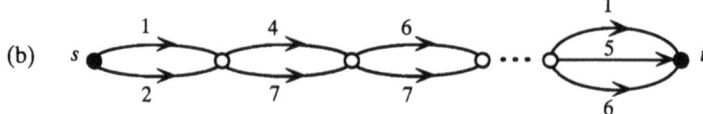

Fig. 4.2 Network interpretation of Esary–Proschan bounds.

edges. Approaches for choosing a set of disjoint cutsets have been investigated by Colbourn (1987a, 1988) and they appear to generate excellent bounds when the p_i are large. In general, it is expected that the disjoint-paths lower bound should be better for small values of the p_i (since we are ignoring the contribution of certain paths), while the disjoint-cutsets upper bound should be better for large values of the p_i.

To illustrate the computation of these bounds, consider the sample network of Fig. 4.1, together with the choice of disjoint paths $\{P_1, P_2\}$ and the choice of disjoint cutsets $\{C_1, C_2\}$. This produces the bounds $DP_L = 1 - (1 - p^2)(1 - p^3)$ and $DC_U = (1 - q^2)^4$, which have been evaluated at selected values of p in Table 4.3. It is seen here that for small values of p the lower bound DP_L is tighter than EP_L, whereas the upper bound EP_U is tighter than DC_U. For large values of p the reverse is true: EP_L is better than DP_L while DC_U is better than EP_U. Thus a combination of these two static bounds can be beneficial.

Several other static bounds have been developed and we briefly indicate certain of these. Shogan (1976, 1977a) used the decomposition of a network into its strongly connected components to obtain an acyclic network whose 'supernodes' correspond to sets of nodes in the original network. For such an acyclic network, the supernodes can be processed in a natural 'topological order' (see Section 4.3.1) and upper bounds on the reliability to each original node can be used sequentially to provide upper bounds for the other nodes. Shanthikumar (1988) obtained an upper bound on two-terminal reliability by identifying a subset $C_1, C_2, \ldots, C_h$ of the s-t cutsets, which, though not disjoint, are sufficiently well structured to allow the efficient calculation of $\Pr(F_1 \cup F_2 \cup \cdots \cup F_h)$. Since a subset of the s-t cutsets are used, we obtain an upper bound on the exact value $R_{st}(G)$. This method can be viewed as a calculation performed on a special type of (cutset) semilattice, a topic that will be developed more fully in Chapter 6.

AboElFotoh and Colbourn (1989) have generalized the idea of disjoint paths and disjoint cutsets to obtain improved bounds on two-terminal reliability. Their basic strategy is to transform the given network G into two other networks G_1 and G_2 in such a way that $R_{st}(G_1) \leq R_{st}(G) \leq R_{st}(G_2)$. Moreover, the transformations are chosen so that G_1 and G_2 are both series–parallel networks. Since the calculation of two-terminal reliability can be carried out efficiently for series–parallel networks, this method produces readily computed lower and upper bounds on the unknown value $R_{st}(G)$.

A number of authors (Ball and Provan 1982; Van Slyke and Frank 1972) have considered the special case when all edge reliabilities assume a common value p, so that the reliability problem is equivalent to determining certain combinatorial coefficients of the reliability polynomial. Bounds on the individual coefficients of this polynomial then provide static bounds on the reliability of the system. Prominent among such bounds are the Kruskal–Katona bounds and the more recent Chari–Provan bounds (Chari and Provan 1989). A fairly extensive development of this topic is provided in the excellent monograph of Colbourn (1987a).

The preceding bounds have all assumed that edges fail in a statistically independent manner. Several authors (Hagstrom and Mak 1986; Lam and Li 1986b; Shier and Spragins 1985) have proposed and analysed models that capture the explicit dependencies among edge failures. An alternative approach (Assous 1986; Zemel 1982) obtains bounds on the reliability of a system with minimal assumptions about the form of the dependence. To illustrate the basic idea of this approach, again suppose that $P_1, P_2, \ldots, P_k$ are the simple s-t paths of G, with E_i being the event that all edges in path P_i operate. Let path P_i contain edges $j = 1, 2, \ldots, w$ and let e_j denote the event that edge j operates. Unlike the previous cases considered, the events e_j are not assumed to be statistically independent. Suppose, however, that a lower bound l_j on $\Pr(e_j)$ is available. Then

$$\Pr(\bar{E}_1 \bar{E}_2 \cdots \bar{E}_k) \le \Pr(\bar{E}_i) = \Pr(\bar{e}_1 \cup \bar{e}_2 \cup \cdots \cup \bar{e}_w) \le \sum_{j=1}^{w} \Pr(\bar{e}_j) \le \sum_{j=1}^{w} (1 - l_j),$$

so that

$$R_{st}(G) = 1 - \Pr(\bar{E}_1 \bar{E}_2 \cdots \bar{E}_k) \ge 1 - \sum_{j=1}^{w} (1 - l_j).$$

Clearly the right-hand side should be maximized to obtain the tightest bound, and thus one seeks a path P_i whose total length, with respect to the edge weights $(1 - l_j)$, is as small as possible. This is simply a shortest s-t path problem on G, and a number of efficient algorithms are available for determining such a path (Dial et al. 1979). In this way we obtain a lower bound on two-terminal reliability even in the presence of dependent edge failures. Assous (1986) and Zemel (1982) have shown that this bound is in a certain well-defined sense the best possible under the given limited information. In an analogous way an upper bound on reliability can be obtained, utilizing the s-t cutsets and known upper bounds on edge operation. While originally intended for use in cases of dependent edge failures, this technique can in fact be used to improve bounds for the independent edge failure case as well (Brecht and Colbourn 1986).

4.2 Algebraic bounds

In this section we develop dynamic bounds on two-terminal reliability based on the algebraic framework presented in Chapter 3. Algebraic techniques have already proved useful in constructing polynomials that bound from below the reliability polynomial (Sections 3.3–3.4). This algebraic perspective will be further exploited here to produce a sequence of upper- and lower-bound polynomials, guaranteed to enclose the exact reliability value for all choices of the edge reliabilities p_i. Moreover, these polynomials will be shown to converge monotonically to the desired reliability polynomial in a finite number of steps.

Recall that a variable x_k can be associated with each edge of a directed network, and we wish to determine the reliability polynomial as a function $R_{st}(\mathbf{x})$ of these edge variables. A fundamental relation is given by eqn (3.12) which expresses this polynomial in terms of certain operations $\oplus$ and $\otimes$ applied to the simple s-t paths. To motivate the subsequent discussion, consider the generalized Jacobi method of Section 3.3, initialized with $z^{(0)} = e_s(1 \oplus A)$. In network terms, $z_t^{(0)}$ corresponds to an algebraic sum over s-t paths having length one (i.e., consisting of a single edge). It is easy to establish inductively that $z_t^{(r-1)}$ corresponds to an algebraic sum over all s-t paths having length at most r. Because the inclusion of nonsimple paths does not affect this algebraic sum, by virtue of property (3.4), $z_t^{(r-1)}$ also represents the algebraic sum over the set $\mathcal{P}_r$ of all *simple* s-t paths of length at most r. Theorem 3.6 then assures us that the following sequence of polynomials $\{L_r(\mathbf{x}): r \geq 1\}$ converges to $R_{st}(\mathbf{x})$:

$$L_r(\mathbf{x}) = \oplus \sum_{P \in \mathcal{P}_r} v(P). \qquad (4.3)$$

Moreover, the particular choice of initial approximation $z^{(0)}$ ensures a nondecreasing sequence that converges in at most $n-1$ iterations, since no simple s-t path ($s \neq t$) has length exceeding $n-1$. We record this fact as

Theorem 4.1. *The sequence* $\{L_r(\mathbf{x})\}$ *satisfies* $L_1(\mathbf{x}) \leqslant L_2(\mathbf{x}) \leqslant \cdots \leqslant L_{n-1}(\mathbf{x}) = R_{st}(\mathbf{x})$.

In the above result, the indicated ordering on polynomials is that defined by (3.7).

When applied to the sample network of Fig. 4.1, eqn (4.3) yields the following sequence of lower-bound polynomials:

$L_1(\mathbf{x}) = 0$

$L_2(\mathbf{x}) = x_1 x_7$

$L_3(\mathbf{x}) = x_1 x_7 \oplus x_2 x_4 x_6$

$L_4(\mathbf{x}) = R_{st}(\mathbf{x}) = x_1 x_7 \oplus x_2 x_4 x_6 \oplus x_1 x_3 x_4 x_6 \oplus x_2 x_4 x_5 x_7.$

In order to define a sequence of upper bounds, it is convenient to renumber the n nodes of $G=(N, E)$ so that $s=1$ and $t=n$. We then consider for each $r \geq 1$ the set $\mathscr{P}^r$ consisting of all simple paths $P: 1 \to i_1 \to i_2 \to \cdots \to i_k \to j$ satisfying $1 \leq i_1, i_2, \ldots, i_k \leq r$ and $j > r$. Relative to this set of paths, the polynomials $U_r(\mathbf{x})$ are defined by:

$$U_r(\mathbf{x}) = \oplus \sum_{P \in \mathscr{P}^r} v(P). \qquad (4.4)$$

We now show that the $U_r(\mathbf{x})$ do indeed provide a sequence of upper-bound polynomials converging to the reliability polynomial.

Theorem 4.2. *The sequence* $\{U_r(\mathbf{x})\}$ *satisfies* $U_1(\mathbf{x}) \geq U_2(\mathbf{x}) \geq \cdots \geq U_{n-1}(\mathbf{x}) = R_{st}(\mathbf{x})$.

Proof. To show that $U_r(\mathbf{x}) \geq U_{r+1}(\mathbf{x})$, consider a simple s-j path $P \in \mathscr{P}^{r+1}$. If $P \in \mathscr{P}^r$ then $v(P)$ already appears in the expression for $U_r(\mathbf{x})$. Suppose, then, that $P \notin \mathscr{P}^r$. Path P must therefore visit node $r+1$ exactly once, so let Q_1 denote the subpath of P from node s to node $r+1$ and let Q_2 denote the subpath of P from $r+1$ to j. By relation (3.9), $v(P) = v(Q_1) \otimes v(Q_2) \leq v(Q_1)$, where $Q_1 \in \mathscr{P}^r$. In either case, $v(P)$ is bounded above by a path value occurring in $U_r(\mathbf{x})$. In addition, there can be path values in $U_r(\mathbf{x})$ not appearing in $U_{r+1}(\mathbf{x})$, but by (3.9) the inclusion of such paths in $U_r(\mathbf{x})$ cannot decrease the overall sum with respect to $\oplus$. Consequently, $U_r(\mathbf{x}) \geq U_{r+1}(\mathbf{x})$ holds for each r and the sequence is nonincreasing. In addition, it is clear from the definition that $\mathscr{P}^{n-1}$ is precisely the set of all simple s-t paths, and so $U_{n-1}(\mathbf{x}) = R_{st}(\mathbf{x})$ follows by (3.12). □

Suppose in the network of Fig. 4.1 that nodes s, a, b, c, t are numbered $1, 2, \ldots, 5$ respectively. Then the upper-bound polynomials calculated from (4.4) are:

$$U_1(\mathbf{x}) = x_1 \oplus x_2$$
$$U_2(\mathbf{x}) = x_2 \oplus x_1 x_3 \oplus x_1 x_7$$
$$U_3(\mathbf{x}) = x_1 x_7 \oplus x_2 x_4 \oplus x_1 x_3 x_4$$
$$U_4(\mathbf{x}) = R_{st}(\mathbf{x}) = x_1 x_7 \oplus x_2 x_4 x_6 \oplus x_1 x_3 x_4 x_6 \oplus x_2 x_4 x_5 x_7.$$

Table 4.4 shows numerical values for these bounds $L_r(p)$ and $U_r(p)$, evaluated at various common edge reliability values p. Notice that useful reliability estimates are obtained for all values of p, unlike the inclusion–exclusion bounds (which can produce invalid probability estimates). In this example, the algebraic bounds are seen to be particularly good for larger values of p, comparable to the best of the Esary–Proschan and disjoint-paths/cutsets bounds.

Table 4.4 Algebraic bounds

p	$L_2(p)$	$L_3(p)$	$R_{st}(G)$	$U_3(p)$	$U_2(p)$	$U_1(p)$
0.0	0.0	0.0	0.0	0.0	0.0	0.0
0.1	0.01000	0.01099	0.01115	0.02071	0.11710	0.19000
0.2	0.04000	0.04768	0.04973	0.08352	0.25760	0.36000
0.3	0.09000	0.11457	0.12251	0.18513	0.40710	0.51000
0.4	0.16000	0.21376	0.23219	0.31744	0.55360	0.64000
0.5	0.25000	0.34375	0.37500	0.46875	0.68750	0.75000
0.6	0.36000	0.49824	0.53971	0.62496	0.80160	0.84000
0.7	0.49000	0.66493	0.70815	0.77077	0.89110	0.91000
0.8	0.64000	0.82432	0.85709	0.89088	0.95360	0.96000
0.9	0.81000	0.94851	0.96163	0.97119	0.98910	0.99000
1.0	1.0	1.0	1.0	1.0	1.0	1.0

The upper bounds produced by (4.4) employ paths $P \in \mathcal{P}^r$ starting at node $s = 1$ and using no intermediate node greater than r. A different set of upper bounds can also be obtained using paths leading into node $t = n$. In particular, we could have defined the set $\mathcal{P}^r$ to consist of all simple paths $P: i \rightarrow i_1 \rightarrow i_2 \rightarrow \cdots \rightarrow i_k \rightarrow n$ satisfying $n - r + 1 \leq i_1, i_2, \ldots, i_k \leq n$, $i < n - r + 1$. Application of eqn (4.4) would then produce another set of upper-bound polynomials, also guaranteed to converge to the reliability polynomial in at most $n - 1$ iterations.

Using either of these definitions together with (4.3), we can, in theory, calculate at each step r a lower-bound polynomial $L_r(\mathbf{x})$ and an upper-bound polynomial $U_r(\mathbf{x})$. While finite convergence to the reliability polynomial $R_{st}(\mathbf{x})$ is guaranteed, it is possible to terminate the procedure after several iterations, yielding absolute bounds on the unknown reliability value. More importantly, it is computationally advantageous to have a mechanism for updating the upper-bound polynomials from iteration to iteration, just as the lower-bound polynomials (4.3) can be updated using the standard Jacobi scheme. The next section discusses various strategies that enable this updating to be carried out effectively.

4.3 Computation of algebraic bounds

In this section we discuss implementation issues arising in calculating algebraic upper and lower bounds. It will be shown that the upper-bound polynomials can in fact be updated in a relatively efficient manner. Moreover, a lower bound will be readily available to match the upper bound calculated at each iteration. Techniques for updating these upper-bound polynomials are discussed in the following subsections.

4.3.1 Acyclic networks

We consider here the case in which the directed network $G = (N, E)$ is acyclic. Even though these represent fairly specialized configurations, such networks regularly occur in project management and scheduling problems, for example. Moreover, the computation of two-terminal reliability for acyclic networks remain #P-complete (Provan 1986c), and so the calculation of bounds is still a useful alternative to the exact calculation of reliability.

Since G is acyclic its nodes can be *topologically numbered* so that if $(i, j) \in E$ then $i < j$. This fact, together with the definition of $\mathscr{P}$, allows the upper-bound polynomials $U_r(\mathbf{x})$ to be updated efficiently. Again, without loss of generality, it can be assumed that $s = 1$ and $t = n$. Consider a path $P \in \mathscr{P}$, where $P: 1 \to i_1 \to i_2 \to \cdots \to i_k \to j$, $j > r$, and the intermediate nodes satisfy

$$1 \leq i_1, i_2, \ldots, i_k \leq r. \tag{4.5}$$

In view of the topological numbering of G, we must actually have $1 < i_1 < i_2 < \cdots < i_k < j$. Now either $P \in \mathscr{P}^{r-1}$ or $P \notin \mathscr{P}^{r-1}$. In the latter case, P must contain node r and furthermore r must be the penultimate node i_k on P. Thus P can be decomposed into the subpath $Q: 1 \to i_1 \to i_2 \to \cdots \to i_k = r$, $Q \in \mathscr{P}^{r-1}$, followed by the edge (r, j). Let $Z_r(j)$ denote the algebraic sum of path values $v(P)$ for $s - j$ paths $P \in \mathscr{P}$ satisfying (4.5). The following simple update formula then incorporates the two cases discussed above:

$$Z_r(j) = Z_{r-1}(j) \oplus [Z_{r-1}(r) \otimes x_{rj}]. \tag{4.6}$$

This update formula directly leads to a method for bounding the two-terminal reliability of acyclic networks, relative to $s = 1$ and $t = n$. Namely, the required upper-bound polynomial $U_r(\mathbf{x})$ can be calculated from

$$U_r(\mathbf{x}) = \oplus \sum_{j=r+1}^{n} Z_r(j) \tag{4.7}$$

and the constituent polynomials $Z_r(j)$ can be updated using (4.6).

To implement this method the following algorithm maintains for each node j a label $L(j)$, representing the current value of the polynomial $Z_r(j)$. The quantities A_{Lr} and A_{Ur} produced at each step of the algorithm are matched lower and upper bounds, respectively, on the exact two-terminal reliability.

Acyclic algorithm. This algorithm generates lower bounds A_{Lr} and upper bounds A_{Ur} on $R_{1n}(\mathbf{x})$ for an acyclic network $G = (N, E)$, whose nodes are assumed to be numbered in topological order.

1. [Initialization]
   $L(1) := 1;\ L(j) := 0$, for $j \neq 1$.

2. [Iterative Step]
   for $r = 1, \ldots, n-1$ do
     for all $(r, j) \in E$ do
       $L(j) := L(j) \oplus [L(r) \otimes x_{rj}]$;
   $A_{Lr} := L(n)$;
   $A_{Ur} := L(r+1) \oplus \cdots \oplus L(n)$.

Each iteration thus produces both upper- and lower-bound polynomials on the unknown reliability polynomial. These polynomial bounds are valid for any specification of the input edge reliabilities p_i. While these bounds will converge in a finite number of iterations, the computational effort increases with each iteration and one would normally terminate the process after relatively few iterations, thus giving guaranteed absolute bounds on the desired reliability.

To illustrate the sequence of bounds developed by this algorithm, consider the acyclic network in Fig. 4.3, whose nodes have been topologically numbered. To calculate the reliability polynomial $R_{16}(\mathbf{x})$, the polynomial labels $L(j)$ on each node are first initialized:

$$L(1) = 1, \quad L(2) = L(3) = L(4) = L(5) = L(6) = 0.$$

Processing of node 1 creates the new labels

$$L(2) = x_3, \quad L(3) = x_4, \quad L(4) = x_2, \quad L(5) = x_1, \quad L(6) = 0.$$

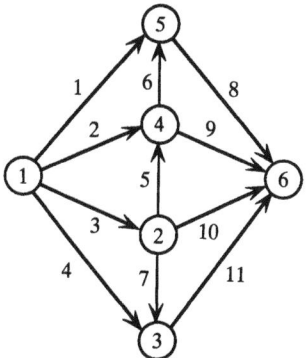

Fig. 4.3 An acyclic network.

The label $L(2) = x_3$ is then propagated, as described in the acyclic algorithm, to nodes 3, 4, and 6, giving

$$L(3) = x_4 \oplus x_3 x_7, \quad L(4) = x_2 \oplus x_3 x_5, \quad L(5) = x_1, \quad L(6) = x_3 x_{10}.$$

Subsequent processing of the remaining nodes in topological order yields the labels:

$$L(4) = x_2 \oplus x_3 x_5, \quad L(5) = x_1, \quad L(6) = x_3 x_{10} \oplus x_4 x_{11} \oplus x_3 x_7 x_{11};$$
$$L(5) = x_1 \oplus x_2 x_6 \oplus x_3 x_5 x_6, \quad L(6) = x_2 x_9 \oplus x_3 x_{10} \oplus x_4 x_{11} \oplus x_3 x_5 x_9 \oplus x_3 x_7 x_{11};$$
$$R_{16}(\mathbf{x}) = L(6) = x_1 x_8 \oplus x_2 x_9 \oplus x_3 x_{10} \oplus x_4 x_{11} \oplus x_2 x_6 x_8 \oplus x_3 x_5 x_9 \oplus x_3 x_7 x_{11} \oplus x_3 x_5 x_6 x_8.$$

Consequently, the lower- and upper-bound polynomials at each iteration are given by:

$$A_{L1} = 0, \quad A_{U1} = x_1 \oplus x_2 \oplus x_3 \oplus x_4,$$
$$A_{L2} = x_3 x_{10}, \quad A_{U2} = x_1 \oplus x_2 \oplus x_4 \oplus x_3 x_5 \oplus x_3 x_7 \oplus x_3 x_{10},$$
$$A_{L3} = x_3 x_{10} \oplus x_4 x_{11} \oplus x_3 x_7 x_{11},$$
$$A_{U3} = x_1 \oplus x_2 \oplus x_3 x_5 \oplus x_3 x_{10} \oplus x_4 x_{11} \oplus x_3 x_7 x_{11},$$
$$A_{L4} = x_2 x_9 \oplus x_3 x_{10} \oplus x_4 x_{11} \oplus x_3 x_5 x_9 \oplus x_3 x_7 x_{11},$$
$$A_{U4} = x_1 \oplus x_2 x_6 \oplus x_2 x_9 \oplus x_3 x_{10} \oplus x_4 x_{11} \oplus x_3 x_5 x_6 \oplus x_3 x_5 x_9 \oplus x_3 x_7 x_{11},$$
$$A_{L5} = A_{U5} = R_{16}(\mathbf{x}).$$

In this example the bounds produced are particularly useful at higher edge reliabilities. For instance, when all $p_i = 0.9$, the bounds and interval widths at iterations 3 and 4 are calculated to be

$$A_{L3} = 0.97119, \quad A_{U3} = 0.99980, \quad \text{width} = 0.02861$$
$$A_{L4} = 0.99532, \quad A_{U4} = 0.99977, \quad \text{width} = 0.00445$$

and, when all $p_i = 0.95$, the bounds and widths are

$$A_{L3} = 0.99264, \quad A_{U3} = 0.99999, \quad \text{width} = 0.00735$$
$$A_{L4} = 0.99939, \quad A_{U4} = 0.99998, \quad \text{width} = 0.00059.$$

It is important to stress that the validity of the update formula given in (4.6) depends crucially on the network being acyclic. When the network contains cycles, the penultimate node on a path $P \in \mathscr{P}^r - \mathscr{P}^{r-1}$ need not equal r and so the simple edge extension implicit in (4.6) is no longer appropriate. Two different methods for dealing with the case of general networks are presented next.

4.3.2 General networks

As discussed in Section 3.2 the calculation of two-terminal reliability can be viewed as the problem of solving a system of linear equations in the operations $\oplus$ and $\otimes$:

$$z = zA \oplus w. \tag{4.8}$$

Here A is the labelled adjacency matrix for the given network G, with z and w being row vectors of polynomials in the algebraic structure $(\mathcal{S}, \oplus, \otimes)$. Of significance here is the minimal solution of (4.8), which by Theorem 3.5 equals wA^*. It is possible to express (4.8) as a pair of systems

$$z = zL \oplus y, \quad y = yU \oplus w, \tag{4.9}$$

which has the same minimal solution z. Moreover, L is a strictly lower triangular matrix, and U is a strictly upper triangular matrix. Thus the given network can be represented by a pair of acyclic networks G_L and G_U, having L and U (respectively) as their labelled adjacency matrices. As shown in Backhouse and Carré (1975), these two matrices can be derived from A via a simple set of transformations, reminiscent of those used to perform an LU decomposition in numerical linear algebra. Define the sequence of matrices $A = A^{(0)}, A^{(1)}, \ldots, A^{(n)}$ by

$$a_{ij}^{(k)} = \begin{cases} a_{ij}^{(k-1)} \oplus [a_{ik}^{(k-1)} \otimes a_{kj}^{(k-1)}] & \text{if } i, j > k \ (i \neq j) \\ a_{ij}^{(k-1)} & \text{otherwise.} \end{cases} \tag{4.10}$$

Then the triangular matrices $L = (l_{ij})$ and $U = (u_{ij})$ are simply given by $l_{ij} = a_{ij}^{(n)}$, $i > j$ and $u_{ij} = a_{ij}^{(n)}$, $i < j$.

Because the networks G_L and G_U are acyclic, Theorem 3.5 shows that the systems in (4.9) have the unique solutions $z = yL^*$ and $y = wU^*$, giving the minimal solution $z = wU^*L^*$. Since (4.8) is known to have the minimal solution wA^* we conclude that $A^* = U^*L^*$. The utility of this decomposition is that we can now apply the results of Section 4.3.1 to obtain upper and lower bounds on reliability in the acyclic networks G_L and G_U. Combining these results then gives bounds on the two-terminal reliability of G. Specifically, the acyclic algorithm in the previous section can be invoked to produce, at each iteration r, a (row) vector $z_r(G_U)$ of upper bounds on the reliabilities from node s. The jth component of this vector represents an upper bound on $R_{sj}(\mathbf{x})$ in G_U. Similarly, the acyclic algorithm yields a (column) vector of upper bounds $z_r(G_L)$ on the reliabilities $R_{it}(\mathbf{x})$ into node t. Relative to the ordering (3.7) we have $z_r(G_U) \geqslant e_s U^*$ and $z_r(G_L) \geqslant L^* e_t^T$, whence $z_r(G) \equiv z_r(G_U) z_r(G_L) \geqslant e_s U^* L^* e_t^T = e_s A^* e_t^T = a_{st}^* = R_{st}(\mathbf{x})$. This shows that by multiplying (with respect to $\otimes$) the iteration r upper bounds for G_L and G_U we obtain a sequence of upper-bound polynomials for G. Because the individual sequences $\{z_r(G_U)\}$ and $\{z_r(G_L)\}$ are nonincreasing, the properties of $\geq$ ensure that $\{z_r(G)\}$ is also nonincreasing. Moreover, these upper bounds on $R_{st}(\mathbf{x})$ are easy to maintain and update. In an analogous fashion, the lower bounds derived from G_L and G_U yield a monotone nondecreasing and convergent sequence of readily updated lower bounds on the reliability polynomial for a general network.

To apply this LU decomposition technique to a given network G, it is useful to number the nodes so that $s = 1$ and $t = 2$. A suitably numbered version of Fig. 4.1 is shown in Fig. 4.4. The figure also shows the labelled adjacency matrix A for the network, as well as the result $A^{(5)}$ of applying the transformations (4.10) to $A = A^{(0)}$. The upper triangular portion of $A^{(5)}$ defines the acyclic network G_U with source node $s = 1$ shown in Fig. 4.4. The lower triangular portion gives the acyclic network G_L with destination node $t = 2$. Table 4.5 displays the sequence of labels $L(j)$ resulting

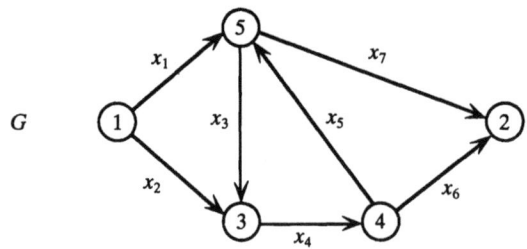

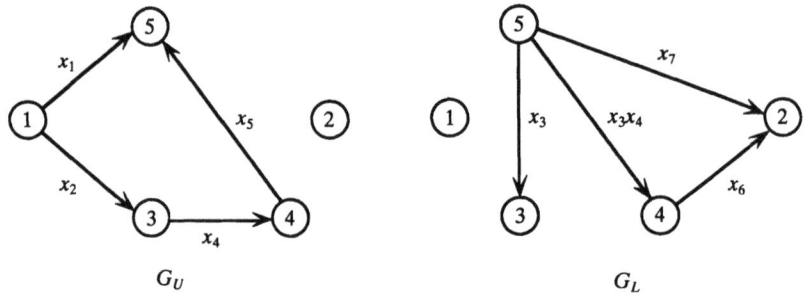

Fig. 4.4 Decomposition of G into G_U and G_L.

Table 4.5 Lower-bound polynomials for the two acyclic networks

	Labels $L(j)$ for G_U				
Iteration	$L(1)$	$L(2)$	$L(3)$	$L(4)$	$L(5)$
0	1	0	0	0	0
1	1	0	x_2	0	x_1
2	1	0	x_2	0	x_1
3	1	0	x_2	$x_2 x_4$	x_1
4	1	0	x_2	$x_2 x_4$	$x_1 \oplus x_2 x_4 x_5$

	Labels $L(j)$ for G_L				
Iteration	$L(1)$	$L(2)$	$L(3)$	$L(4)$	$L(5)$
0	0	1	0	0	0
1	0	1	0	0	0
2	0	1	0	x_6	x_7
3	0	1	0	x_6	x_7
4	0	1	0	x_6	$x_7 \oplus x_3 x_4 x_6$

from applying the acyclic algorithm to these two networks. By forming the product, with respect to $\otimes$, of the corresponding vectors of lower bounds in Table 4.5, we obtain the following lower-bound polynomials $L_r(\mathbf{x})$ for $R_{st}(\mathbf{x})$:

$L_1(\mathbf{x}) = 0,$

$L_2(\mathbf{x}) = x_1 x_7,$

$L_3(\mathbf{x}) = x_1 x_7 \oplus x_2 x_4 x_6,$

$L_4(\mathbf{x}) = R_{st}(\mathbf{x}) = x_2 x_4 x_6 \oplus (x_1 \oplus x_2 x_4 x_5)(x_7 \oplus x_3 x_4 x_6)$
$= x_1 x_7 \oplus x_2 x_4 x_6 \oplus x_1 x_3 x_4 x_6 \oplus x_2 x_4 x_5 x_7.$

In order to generate the upper-bound polynomials, we first associate with any node j in G_U the following sum of labels produced by the acyclic algorithm at iteration r:

$$S(j) = L(j) \oplus \sum_{k=r+1}^{j-1} L(k).$$

Here the indicated summation is taken to be 0 if the associated index set is empty. The quantity $S(j)$ corresponds to the upper bound produced by the acyclic algorithm relative to destination node j. The same formula is used to produce analogous labels $S(j)$ in G_L. These new labels are shown in Table 4.6 for both networks G_U and G_L. Then an upper-bound polynomial $U_r(\mathbf{x})$ for G is obtained by multiplying, again with respect to $\otimes$, the corresponding vectors of upper bounds $S(j)$ in Table 4.6, yielding

$$U_1(\mathbf{x}) = x_1 \oplus x_2,$$

$$U_2(\mathbf{x}) = x_2 x_6 \oplus (x_1 \oplus x_2)(x_6 \oplus x_7) = x_1 x_6 \oplus x_1 x_7 \oplus x_2 x_6 \oplus x_2 x_7,$$

$$U_3(\mathbf{x}) = x_2 x_4 x_6 \oplus (x_1 \oplus x_2 x_4)(x_6 \oplus x_7) = x_1 x_6 \oplus x_1 x_7 \oplus x_2 x_4 x_6 \oplus x_2 x_4 x_7,$$

$$U_4(\mathbf{x}) = R_{st}(\mathbf{x}) = x_2 x_4 x_6 \oplus (x_1 \oplus x_2 x_4 x_5)(x_7 \oplus x_3 x_4 x_6)$$
$$= x_1 x_7 \oplus x_2 x_4 x_6 \oplus x_1 x_3 x_4 x_6 \oplus x_2 x_4 x_5 x_7.$$

An alternative to the LU decomposition approach is to develop an appropriate update formula for the polynomials $Z_r(j)$, analogous to (4.6) for acyclic networks, which will be valid for general networks. Consequently, we focus on the structure of the set of paths $\mathscr{P}^r$. Rather than calculating this set of paths independently for each iteration r, we pursue an updating scheme that allows the set of paths at iteration $r+1$ to be deduced from that at iteration r. To this end, define, for $2 \le r \le n-1$, the set $\mathscr{P}^r(i, j)$ containing all simple paths $P: i \to i_1 \to i_2 \to \cdots \to i_k \to j$

Table 4.6 Upper-bound polynomials for the two acyclic networks

	Labels $S(j)$ for G_U				
Iteration	$S(1)$	$S(2)$	$S(3)$	$S(4)$	$S(5)$
0	1	1	1	1	1
1	1	0	x_2	x_2	$x_1 \oplus x_2$
2	1	0	x_2	x_2	$x_1 \oplus x_2$
3	1	0	x_2	$x_2 x_4$	$x_1 \oplus x_2 x_4$
4	1	0	x_2	$x_2 x_4$	$x_1 \oplus x_2 x_4 x_5$

	Labels $S(j)$ for G_L				
Iteration	$S(1)$	$S(2)$	$S(3)$	$S(4)$	$S(5)$
0	0	1	1	1	1
1	0	1	1	1	1
2	0	1	0	x_6	$x_6 \oplus x_7$
3	0	1	0	x_6	$x_6 \oplus x_7$
4	0	1	0	x_6	$x_7 \oplus x_3 x_4 x_6$

that satisfy relation (4.5). By definition, $\mathscr{P}^1(i, j)$ consists of the edge (i, j), if it exists in the network G. Also, let $Z_r(i, j)$ denote the polynomial obtained by adding, with respect to $\oplus$, all path values $v(P)$ for $P \in \mathscr{P}^r(i, j)$. Assuming that $s = 1$ and $t = n$, then $Z_r(j) = Z_r(1, j)$ and by (4.7) the required upper-bound polynomial is given by

$$U_r(\mathbf{x}) = \oplus \sum_{j=r+1}^{n} Z_r(1, j). \tag{4.11}$$

The following result gives a simple update formula for the polynomials $Z_r(i, j)$, similar to the update relation encountered in Floyd's shortest-path algorithm (Floyd 1962).

Theorem 4.3. *For $r = 2, \ldots, n - 1$*

$$Z_r(i, j) = Z_{r-1}(i, j) \oplus [Z_{r-1}(i, r) \otimes Z_{r-1}(r, j)]. \tag{4.12}$$

Proof. Any path $P \in \mathscr{P}^r(i, j)$ either uses node r or does not. If it does not, then $v(P)$ is included in $Z_{r-1}(i, j)$. Otherwise P can be decomposed into subpaths Q_1 and Q_2, with Q_1 a simple i-r path and Q_2 a simple r-j path. Since $P \in \mathscr{P}^r(i, j)$ both these subpaths contain no intermediate node larger than $r - 1$. As a result, $v(P) = v(Q_1) \otimes v(Q_2)$ occurs in the product $Z_{r-1}(i, r) \otimes Z_{r-1}(r, j)$. Conversely, any product of this form either corresponds to a simple i-j path satisfying (4.5) or contains a simple path $P \in \mathscr{P}^{r-1}(i, j)$. In the latter case, by property (3.4), this product will then be 'absorbed' into a path value occurring in $Z_{r-1}(i, j)$. This establishes the formula (4.12). □

By virtue of Theorem 4.3, the set of polynomials $Z_r(i, j)$ at iteration r can be readily updated, based on the corresponding polynomials at the previous iteration. It is easy to establish by induction that not all such polynomials need to be maintained at each iteration. Indeed, at any iteration $r = 2, \ldots, n - 1$ one need only compute the polynomials $Z_r(i, j)$ for the following ranges of i and j:

$$Z_r(1, j) \quad \text{for } j = r + 1, \ldots, n,$$

$$Z_r(i, j) \quad \text{for } i = r + 1, \ldots, n - 1; \quad j = r + 1, \ldots, n; \quad i \neq j.$$

The required information will then be available for computing the upper-bound polynomial $U_r(\mathbf{x})$ using (4.11). A convenient lower-bound polynomial is also readily available at each iteration r, namely

$$L_r(\mathbf{x}) = Z_r(1, n). \tag{4.13}$$

It is of interest to point out that the coefficients of every lower-bound polynomial $L_r(\mathbf{x})$ in (4.13) and every upper-bound polynomial $U_r(\mathbf{x})$ in (4.11), when fully expanded and simplified, assume only the values 0 or

±1. This result is similar in spirit to Theorem 2.1, which states that the (fully reduced) coefficients of $R_{st}(\mathbf{x})$ must indeed be of this form. It is not true, in general, that any collection of paths in a network will yield bounding polynomials with this property. However, the special nature of the paths embodied in the upper- and lower-bound polynomials allows this conclusion to be derived from Theorem 2.1.

To see this, consider, first, the lower-bound polynomial $L_r(\mathbf{x}) = Z_r(1, n)$, which involves the algebraic sum of path values over all paths $P \in \mathcal{P}(1, n)$. Since the intermediate nodes of P must be in the set $R = \{1, 2, \ldots, r\}$, such a path is also a 1-n path in the subnetwork G_r of G induced by the nodes $R \cup \{n\}$. That is to say, G_r has the node set $R \cup \{n\}$ and contains all edges of G joining two nodes in this set. Because of the one-to-one correspondence between 1-n paths in $\mathcal{P}(1, n)$ and simple 1-n paths in G_r, it is clear that $L_r(\mathbf{x})$ is just the reliability polynomial $R_{1n}(\mathbf{x}; G_r)$. (Here we have made explicit the network underlying the reliability calculations.) From Theorem 2.1 it follows that the coefficients of $L_r(\mathbf{x})$ must equal 0 or ±1. Moreover, the nonzero coefficients in $L_r(\mathbf{x})$ again correspond to the acyclic relevant subnetworks in G_r, with the algebraic signs given by Theorem 2.1.

Likewise, we can analyse the coefficients appearing in the upper-bound polynomials $U_r(\mathbf{x})$ generated at each iteration. Each path value represented in (4.11) derives from a path of $\mathcal{P}(1, j)$ for $j > r$, and so has intermediate nodes in R. Construct from G the network G' in which all nodes $j > r$ are coalesced into a supernode α. That is, G' has node set $R \cup \{\alpha\}$ and contains all edges (u, v) of G with $u, v \in R$ and has edges (u, α) replacing edges (u, v) of G with $u \in R$ and $v \notin R$. We can see that there is a one-to-one correspondence between paths in $\mathcal{P}(1, j)$, $j > r$, and simple 1-α paths in G', showing that $U_r(\mathbf{x}) = R_{1\alpha}(\mathbf{x}; G')$. Theorem 2.1 can again be invoked to show that the coefficients of $U_r(\mathbf{x})$ must be 0 or ±1, with the appropriate ±1 signs reserved for the acyclic relevant subnetworks of G'.

4.4 Chapter notes

This chapter has briefly surveyed a number of methods for bounding the reliability of networks, particularly in the case of two-terminal directed networks. A wealth of powerful combinatorial tools can be applied to study other network reliability problems, notably the all-terminal reliability in directed and undirected networks. Van Slyke and Frank (1972) apply certain results of Kruskal (1963) and Katona (1966) to obtain reliability bounds for coherent systems, whereas Ball and Provan (1982, 1983) develop improved bounds for shellable systems (which include the all-terminal reliability problem). A detailed explanation of the combinatorics underlying these various bounding methods can be found in Colbourn (1987a). In addition, Provan (1986a, b) gives a nice exposition of the important combinatorial ideas that are used in deriving certain of these bounds.

The algebraic bounding approaches developed in Sections 4.2 and 4.3 address the two-terminal problem and involve symbolic manipulations with reliability polynomials. Shier and Whited (1988) derive a convergent sequence of upper-bound polynomials for the case of acyclic networks. Computational results are also presented using these upper bounds together with companion lower bounds. Shier and Liu (1989) provide a different type of upper bound, applicable to general networks, and illustrate the efficacy of the bounds on certain standard communication network topologies.

Recent work has also been carried out in applying sophisticated simulation methodologies to the calculation of network reliability. Karp and Luby (1983) use a sampling technique based on cutsets in the network to estimate reliability. Fishman (1986a, b; 1987) has developed and compared several Monte Carlo sampling plans for estimating reliability. Typically, these methods depend upon the availability of known upper and lower bounds on the reliability parameters to be estimated, and in this sense the bounding techniques of this chapter can be quite practical when used in conjunction with such simulation methods. Nel and Colbourn (1990) have used Monte Carlo techniques to sample certain spanning trees of a network, thus producing upper and lower bounds on the all-terminal reliability for the network.

5

Enumeration of paths and cutsets

Several exact methods for calculating network reliability, such as the inclusion–exclusion and disjoint-products techniques, require that the simple s-t paths or the s-t cutsets be first enumerated. Likewise, a number of methods for obtaining bounds on the exact reliability (Section 4.1) are predicated on a knowledge of the s-t paths and cutsets. It will be seen in the next chapter that certain 'pseudopolynomial' algorithms for calculating reliability also require that such objects be generated in advance. Consequently, this chapter discusses various methods for enumerating the paths and cutsets of a directed network. Of particular interest will be algebraic enumeration methods that parallel and extend the algebraic techniques presented in previous chapters.

In a given network $G = (N, E)$ having $n = |N|$ nodes and $m = |E|$ edges, the number of paths and cutsets can grow rapidly with n and m. Specifically, the number of simple s-t paths in a connected network can be as large as 2^{m-n+1}, whereas the number of s-t cutsets can be as large as 2^{n-2}. Therefore, all enumeration techniques possess a computational complexity that, in the worst case, grows exponentially with the network size. Since, in practice, networks tend to be relatively sparse and of limited connectivity, the actual number of paths or cutsets may in fact remain of reasonable size. Accordingly, it is desirable to have generation methods whose complexity grows only polynomially with the number of paths or the number of cutsets. Our discussion first turns to methods for generating the simple paths of G, which is conceptually and practically more straightforward than the generation of cutsets.

5.1 Path-enumeration methods

A well-studied technique for exploring the nodes and edges of a given network is called *depth-first search*. Essentially it starts at some root node and always moves forward along an unexplored edge until no further progress is possible. At that point the procedure backtracks to the last explored node and resumes the search from that point. This technique can be easily adapted to yield a method for enumerating all simple s-t paths of a directed network $G = (N, E)$. Initially the source (root) node s is 'labeled' and as each new node is encountered it, too, is labelled. Once the destination node t has been reached, the labelled nodes (in the order encountered) constitute an s-t path. This method can be compactly expressed using the recursive procedure DFS given below. The enumeration of all s-t paths is accomplished by invoking DFS(s).

Procedure DFS(i). Given a network $G = (N, E)$, this procedure generates the simple paths from node i to node t.

label node i;
for $(i, j) \in E$ do
   if j is not labelled then
     if $j \neq t$ then DFS(j)
     else output new s-t path;
unlabel node i.

While this procedure is easy to implement and is often quite efficient in practice, there are networks with relatively few paths in which DFS requires an exorbitant amount of time. Fig. 5.1 shows a network on $n+2$ nodes containing one simple s-t path: $s \to 1 \to t$. At node 1 is attached a subnetwork H which is completely connected: any two distinct nodes of H are connected by an edge. Once the depth-first search enters H, a great deal of time is spent recursively exploring this subnetwork without producing any simple s-t paths. Such examples serve to demonstrate that algorithm DFS can require an amount of time that is exponential in the number of s-t paths. In order to remedy this defect, modifications can be made to the basic DFS approach so that the effort spent per generated path is bounded. The basic idea is to look ahead from the node j just reached in the depth-first search to see if node t can, in fact, be reached by a path disjoint from the current path. The resulting algorithm has been shown by Read and Tarjan (1975) to have complexity $O(n + m + mp_{st})$, where p_{st} is the number of simple s-t paths. Another algorithm, based on the same general principle and also having polynomially bounded complexity in p_{st}, is presented by Colbourn (1987a).

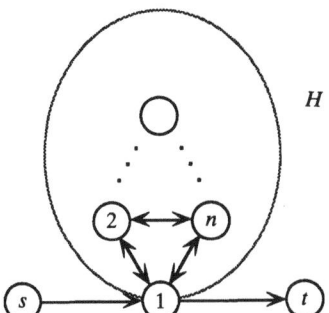

Fig. 5.1 A network on which DFS is inefficient.

While the focus here has been on generating s-t paths, there are other instances in which one would like to generate all the paths emanating from a given source node s: all simple s-j paths for $j \in N$. For example, the generation of such paths has been useful in developing procedures for ranking athletic teams (Perticone 1983), in analysing the structure of computer programs (Fosdick and Osterweil 1976), and in carrying out heuristic search procedures (Nilsson 1980). This single-source generation problem can be approached by suitably modifying the algorithm DFS given above, so that all paths from node s are produced by invoking DFSALL(s). The recursive procedure DFSALL (described below) has complexity that is polynomially bounded in the number of paths originating from the source node.

Procedure DFSALL(i). Given a network $G = (N, E)$, this procedure generates all simple paths from node i.

label node i;
output current path;
for $(i, j) \in E$ do
  if j is not labelled then DFS(j);
unlabel node i.

An algebraic approach can also be used to generate all simple paths from node s. This should not be surprising, since the reliability methods discussed in Chapter 3 implicitly generate the s-t paths in the course of their calculations. To present this algebraic approach, consider an *alphabet* of symbols $\mathcal{A} = \{a, b, c, \ldots\}$ and *words* α over this alphabet (each constituting a string of symbols from $\mathcal{A}$). The concatenation of two words α and β is denoted $\alpha \circ \beta$; the *empty* word λ satisfies $\alpha \circ \lambda = \lambda \circ \alpha = \alpha$ for all α. A *language* X over $\mathcal{A}$ is just a set of words. The *reduced language* $r(X)$ is obtained from X by removing any word that properly contains another word already present in X. For example, the word $\alpha = dabcab$ properly contains the word $\beta = aca$. A language X is called *reduced* if $r(X) = X$ and we will be interested in the set $\mathcal{S}$ of all reduced languages over $\mathcal{A}$. Notice that the empty set $\varnothing$ and the singleton set $\{\lambda\}$ are reduced languages.

Now define two operations $\oplus$ and $\otimes$ on reduced languages $X, Y \in \mathcal{S}$ using

$$X \oplus Y = r(X \cup Y), \tag{5.1}$$

$$X \otimes Y = \{\alpha \circ \beta : \alpha \in X, \beta \in Y\}. \tag{5.2}$$

Then $(\mathcal{S}, \oplus, \otimes)$ is an algebraic structure satisfying many of the properties (3.1)–(3.6) discussed earlier with regard to reliability polynomials. The additive identity 0 for this system is $\varnothing$ and the multiplicative identity 1 is $\{\lambda\}$. The following properties are seen to hold for reduced languages $X, Y, Z \in \mathcal{S}$.

$$X \oplus X = X \tag{5.3}$$

$$X \oplus Y = Y \oplus X \tag{5.4}$$

$$X \oplus (Y \oplus Z) = (X \oplus Y) \oplus Z \qquad X \otimes (Y \otimes Z) = (X \otimes Y) \otimes Z \tag{5.5}$$

$$X \oplus (X \otimes Y) = X \tag{5.6}$$

$$X \otimes (Y \oplus Z) = (X \otimes Y) \oplus (X \otimes Z)$$

$$(X \oplus Y) \otimes Z = (X \otimes Z) \oplus (Y \otimes Z) \tag{5.7}$$

$$X \oplus 0 = X \qquad X \otimes 1 = 1 \otimes X = X \tag{5.8}$$

$$X \otimes 0 = 0 \otimes X = 0. \tag{5.9}$$

Notice that the idempotent, commutative, and absorptive properties do not hold for the multiplicative operation (5.2).

Suppose that each edge e of a given network G is labelled with the reduced language $\{a(e)\}$, where $a(e)$ is a unique symbol from the alphabet. Then the value $v(P)$ of any path $P = [e_1, e_2, \ldots, e_k]$ is given by the product

$$v(P) = \{a(e_1)\} \otimes \{a(e_2)\} \otimes \cdots \otimes \{a(e_k)\} = \{a(e_1)a(e_2)\cdots a(e_k)\},$$

which is just the language corresponding to the word defined by P. Moreover, the optimal value a_{ij}^* defined by

$$a_{ij}^* = \bigoplus_{P \in \mathcal{P}_{ij}} v(P)$$

now has the interpretation as the language whose words correspond to all simple i-j paths. (Recall that $\mathcal{P}_{ij}$ designates the set of simple paths extending from node i to node j.) Thus the calculation of optimal values in this new algebraic structure is tantamount to the enumeration of simple paths. Several iterative techniques were presented in Chapter 3 for producing simultaneously all optimal values a_{sj}^* for $j \in N$. These techniques remain valid even when the weaker set of properties (5.3)–(5.9) is assumed, and therefore they can be applied in the present situation to provide a listing of all s-j paths in G.

There are also occasions when generation of the simple paths between all pairs of nodes is useful. Rather than invoking DFSALL(s) in turn for each node $s \in N$ or using the iterative techniques of Chapter 3 for each node $s \in N$, alternative methods exist that explicitly take into account the interrelationships among paths having different source nodes s. These algebraic techniques correspond to *direct* (rather than iterative) methods for solving the system $z = zA \oplus w$, where A is the labelled adjacency matrix for G. One such direct method for solving ordinary systems of linear equations is the Gauss–Jordan method, and it can be appropriately modified (Carré 1971) to calculate the entire matrix of optimal values A^*. The following algorithm summarizes the steps of this *generalized Gauss–Jordan method*.

Gauss–Jordan algorithm. Given a network $G = (N, E)$, this algorithm transforms the $n \times n$ labelled adjacency matrix $A = (a_{ij})$ into the matrix A^*.

for $k = 1, \ldots, n$ do
  for $i = 1, \ldots, n$ do
    for $j = 1, \ldots, n$ do
      if $i, j \neq k$ then $a_{ij} := a_{ij} \oplus (a_{ik} \otimes a_{kj})$.

Relative to the operations $\oplus$ and $\otimes$ defined in (5.1)–(5.2), the values a_{ij} obtained upon termination of this algorithm correspond to the

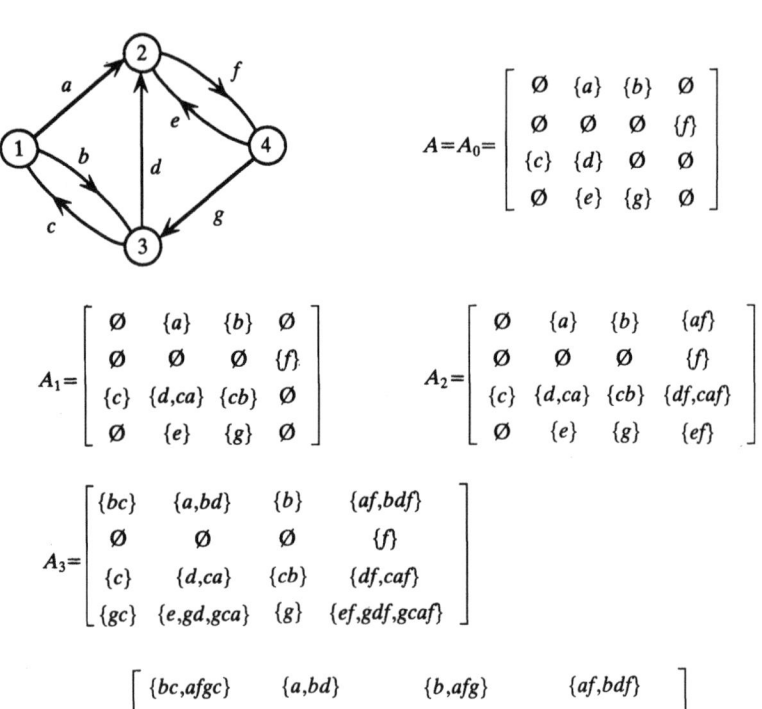

Fig. 5.2 Application of the Gauss–Jordan method for calculating A^*.

required i-j simple paths. This technique for path generation has been discussed by Murchland (1965) and Benzaken (1968). By appropriate choices of the operations $\oplus$ and $\otimes$, a variety of other optimal-path problems can be solved by the Gauss–Jordan algorithm, including the calculation of shortest paths, maximum capacity paths, and the transitive closure of a binary relation (Floyd 1962; Robert and Ferland 1968; Roy 1959; Warshall 1962). A more general version of this algorithm has also been used to study finite automata and regular expressions (Kleene 1956; McNaughton and Yamada 1960), kth shortest paths (Minieka and Shier 1973), and global flow analysis (Tarjan 1981a).

To illustrate the Gauss–Jordan method for generating all simple paths, consider the network G shown in Fig. 5.2. The edges of G have been labelled with symbols from the alphabet $\mathcal{A} = \{a, b, \ldots, g\}$. Each entry of the labelled adjacency matrix A represents a language containing the appropriate edge symbol, or is $\emptyset$ in the case of nonexistent edges. Applying the Gauss–Jordan method to $A = A_0$ results in the sequence of matrices $A_1, \ldots, A_4 = A^*$ shown in Fig. 5.2. Examination of the $(4, 2)$ entry of A^* indicates that the simple paths from node 4 to node 2 are e, gd, and gca; the $(3, 3)$ entry of A^* reveals that the simple cycles on node 3 are cb, dfg, and $cafg$.

5.2 Cutset-enumeration methods

This section studies the generation of cutsets in a directed network G. While the enumeration of cutsets is, in general, more complicated than the enumeration of simple paths, there are several advantages to using them in reliability analysis. If the number of s-t cutsets is smaller than the number of simple s-t paths, then cutset-based procedures for calculating reliability would be more efficient. Moreover, as seen in Section 4.1, the best bounding techniques for large edge reliabilities (the commonly occurring case) are typically those that employ the s-t cutsets rather than the s-t paths. In addition, as described in the next chapter, it is possible to devise an exact algorithm for calculating the two-terminal reliability of G that is polynomial in the number of s-t cutsets. It is unlikely, however, that a comparable algorithm exists that is polynomial in the number of s-t paths (Provan and Ball 1984).

We begin by investigating the structure of s-t cutsets in a directed network. Recall that an s-t (edge) disconnecting set of $G = (N, E)$ is a set $S \subseteq E$ such that $G - S$ contains no s-t path. An s-t cutset S is a minimal such disconnecting set: i.e., no proper subset of S is an s-t disconnecting set. To characterize the cutsets in a more useful way, it is convenient to introduce the concept of a *cut*. If X is a nonempty proper subset of N with $s \in X$ and $t \in \bar{X}$, then the cut $\langle X, \bar{X} \rangle$ denotes the set of all edges

$(i, j) \in E$, with $i \in X$ and $j \in \bar{X}$. Note that removing the edges of $\langle X, \bar{X} \rangle$ separates node s from node t, so that each cut is an s-t disconnecting set but it need not be minimal. For example, in the network of Fig. 5.3, using $X = \{1, 2, 4\}$ produces the cut $\langle X, \bar{X} \rangle = \{b, c, f, g\}$ separating node 1 from node 5. This is not a cutset since $\{b, c, f\}$ also separates node 1 from node 5. The set of edges $\{b, c, f\}$ is actually a 1-5 cutset, and furthermore it can be realized as the cut $\langle X, \bar{X} \rangle$ with $X = \{1, 2\}$.

The following result shows that every cutset is necessarily a cut, and it characterizes exactly which cuts are cutsets. In the statement of the theorem, the *open neighbourhood* $\Gamma(X)$ of X is defined by $\Gamma(X) = \{j \in \bar{X} : (i, j) \in E \text{ for some } i \in X\}$.

Theorem 5.1. $S \subseteq E$ is an s-t cutset if and only if $S = \langle X, \bar{X} \rangle$ for some $X \subseteq N$ satisfying
  (1) $s \in X$ and $t \in \bar{X}$;
  (2) there is a directed path in X from s to each $i \in X$, $i \neq s$;
  (3) there is a directed path in $\bar{X}$ from each $j \in \Gamma(X)$ to t, $j \neq t$.

Proof. ($\Rightarrow$) Suppose that S is an s-t cutset of G, and let X consist of node s together with all nodes accessible (by a directed path) from s in $G - S$. Certainly $s \in X$ and condition (2) is satisfied. Also, $t \in \bar{X}$ since S is an s-t cutset. Now for any edge $e = (i, j) \in S$, $S - e$ does not separate s from t so that edge e must lie on an s-t path in $G - (S - e)$. This means there is an s-i path in $G - S$, whence $i \in X$. Moreover, there exists a j-t path in $G - S$. This implies that $j \in \bar{X}$; for if $j \in X$ we would have an s-j path in $G - S$ and hence an s-t path in $G - S$. Consequently, $e \in \langle X, \bar{X} \rangle$ and $S \subseteq \langle X, \bar{X} \rangle$. On the other hand, if $e = (i, j) \in \langle X, \bar{X} \rangle$, then $e \in S$; otherwise there would be an s-j path in $G - S$, giving $j \in X$, a contradiction. We have thus shown that $S = \langle X, \bar{X} \rangle$. Finally, let $j \in \Gamma(X)$, so that there is some $i \in X$ with $e = (i, j) \in \langle X, \bar{X} \rangle = S$. As before, since S is a minimal disconnecting set, there must be an s-t path in $G - (S - e)$ using edge e and thus a j-t path in $G - S$, whence condition (3) is satisfied.

($\Leftarrow$) Let $S = \langle X, \bar{X} \rangle$ satisfy conditions 1–3. Since $s \in X$ and $t \in \bar{X}$, S is

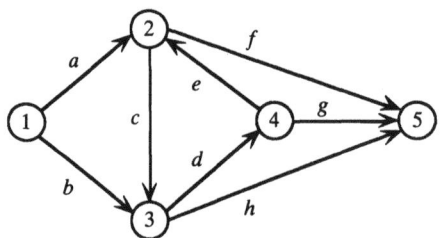

Fig. 5.3 An illustrative network for cutset enumeration.

an s-t disconnecting set. To show that it is a minimal disconnecting set, suppose that $S - e$ is a disconnecting set with $e = (i, j) \in S$. But, by (2) and (3), there is an s-i path in $G - S$ as well as a j-t path in $G - S$. This yields an s-t path in $G - (S - e)$, a contradiction. □

Recall that an undirected network G can be modelled as a directed network G' by replacing each undirected edge $[i, j]$ by two oppositely directed edges (i, j) and (j, i). Generating the s-t cutsets in the directed network G' will then produce the s-t cutsets for the original network G. In view of this equivalence, the conditions of Theorem 5.1 simplify considerably for undirected networks G. Namely, each cutset of G is a cut $\langle X, \bar{X} \rangle$ in which the (induced) subnetworks on X and on $\bar{X}$ are both connected. This observation has led to several algorithms for the generation of s-t cutsets in undirected networks (Abel and Bicker 1982; Bellmore and Jensen 1970; Tsukiyama et al. 1980). The most efficient of these is due to Tsukiyama et al. (1980) and its worst-case time complexity is given by $O((n + m)c_{st})$, where c_{st} is the number of s-t cutsets in G. The Tsukiyama approach does not obviously generalize to the directed case. However, using the concept of 'dominators' in directed graphs, it is possible to construct an algorithm having a worst-case complexity that is polynomial in the number of directed s-t cutsets (Provan and Shier 1990).

Our attention now turns to an algebraic approach that can be used for generating the cutsets of a directed network. To this end, we consider an alphabet of symbols $\mathcal{A} = \{a, b, c, \ldots\}$ and sets X of subsets of symbols from $\mathcal{A}$. A *reduced set* $r(X)$ can be derived from X by removing any subset that properly contains another subset in X. Set X is said to be *reduced* if $r(X) = X$, and we consider the collection $\mathcal{S}$ of all reduced sets over $\mathcal{A}$. For $X, Y \in \mathcal{S}$ we define the two operations $\oplus$ and $\otimes$ using

$$X \oplus Y = r(\{\alpha \cup \beta : \alpha \in X, \beta \in Y\}), \qquad (5.10)$$

$$X \otimes Y = r(X \cup Y). \qquad (5.11)$$

Then it is straightforward to verify that the algebraic structure $(\mathcal{S}, \oplus, \otimes)$ satisfies properties (3.1)–(3.6). The additive identity for this system is $\{\emptyset\}$ and the multiplicative identity is $\emptyset$.

This algebraic structure turns out to be appropriate for describing the cutsets of a network (Martelli 1974). The reasoning here is based on the fact that an i-j cutset 'blocks' all i-j paths in G: it is a minimal set of edges that intersects all i-j paths. This implicit description of cutsets leads naturally to an algebraic formulation in terms of the simple paths. Suppose that a symbol $a(e)$ from $\mathcal{A}$ is associated with each edge e of a given network. We will label this edge with the reduced set $\{\{a(e)\}\}$. Consider a simple i-j path P and define its value $v(P)$ as the product, with respect to $\otimes$, of the path labels occurring along P. Interpreted in terms of the operation $\otimes$ defined in (5.11), $v(P)$ is simply the collection of sets $\{a(e)\}$ corresponding to the edges of P. An i-j cutset must intersect at least one of the edges of P, so each subset present in $v(P)$ indicates a possible choice of a cutset edge. Since an i-j cutset must intersect all i-j paths, the collection of all such cutsets is given by an appropriate sum, relative to (5.10), over all the simple i-j paths:

$$a_{ij}^* = \oplus \sum_{P \in \mathscr{P}_{ij}} v(P).$$

Thus a_{ij}^* correctly enumerates all cutsets separating node i from node j. Once this algebraic association has been made, the various iterative and direct methods mentioned in Section 5.1 can be applied here as well. One iterative technique for simultaneously calculating all s-j cutsets ($j \in N$) has been investigated by Shier and Whited (1986). The computational evidence presented there indicates that the empirical complexity of the algorithm grows approximately with the square of the number of cutsets generated. An alternative approach for generating the i-j cutsets simultaneously for all nodes i and j can be based on the Gauss–Jordan algorithm of Section 5.1. Martelli (1976) discusses this approach, as well as various computational refinements for enhancing the efficiency of the algebraic calculations.

To illustrate the application of this algebraic structure to cutset enumeration, consider the directed network of Fig. 5.3 whose edges are identified by the symbols $a, b, \ldots, h$. We adapt the list-directed iterative approach of Section 3.4 to calculate all s-j cutsets relative to the source node $s = 1$. In describing this algorithm, it will be convenient to represent sets of edge symbols by the corresponding string, so that $\{a, b, d\}$ will be abbreviated as abd. At any step of the algorithm the label $L(j)$ on node j will indicate a collection of s-j cutsets or edge sets that can potentially be expanded into s-j cutsets. Initially all nodes $j \neq 1$ receive the label $L(j) = \{\emptyset\}$ and node 1 receives the label $\emptyset$. The list of nodes to be processed is initialized with $\mathscr{L} = [1]$. Nodes i are successively removed from $\mathscr{L}$ and the labels of adjacent nodes j are updated using $L(j) := L(j) \oplus [L(i) \otimes \{a(e)\}]$, where $e = (i, j)$. In our example, node 1 is first removed from $\mathscr{L}$ and this results in the updated labels $L(2) = \{\emptyset\} \oplus [L(1) \otimes \{a\}] = \{a\}$ and $L(3) = \{\emptyset\} \oplus [L(1) \otimes \{b\}] = \{b\}$, giving $\mathscr{L} = [2, 3]$. At the next step, node 2 is removed from $\mathscr{L}$ and processed, giving $L(3) = \{b\} \oplus [L(2) \otimes \{c\}] = \{ab, bc\}$, $L(5) = \{\emptyset\} \oplus [L(2) \otimes \{f\}] = \{a, f\}$, and $\mathscr{L} = [3, 5]$. Then node 3 is removed from $\mathscr{L}$ and processed: $L(4) = \{\emptyset\} \oplus [L(3) \otimes \{d\}] = \{ab, bc, d\}$, $L(5) = \{a, f\} \oplus [L(3) \otimes \{h\}] = \{ab, ah, bcf, fh\}$, and $\mathscr{L} = [5, 4]$. The procedure is continued until, at the end of iteration 8, the list $\mathscr{L}$ becomes empty. At this point the labels on nodes j are:

$$L(1) = \varnothing, \quad L(2) = \{ab, ad, ae\}, \quad L(3) = \{ab, bc\},$$
$$L(4) = \{ab, bc, d\}, \quad L(5) = \{ab, adh, aegh, bcf, dfh, fgh\},$$

and these represent all minimal 1-j cutsets in the network.

5.3 Chapter notes

The depth-first search technique is a standard algorithmic tool; further details on this method and its properties are provided in Aho *et al.* (1974). The algebraic formulation of simple path enumeration given in Section 5.1 can be found in Carré (1979) and Zimmermann (1981). An alternative formulation of the path-enumeration problem, using 'Latin multiplication', is discussed by Gondran and Minoux (1984). Tarjan (1981*b*) describes an efficient algorithm for producing a regular expression that encodes all paths emanating from a single node in the network. The use of Gauss–Jordan elimination to compute the closure matrix A^* has a long history, extending from its use in determining the transitive closure of a relation (Roy 1959) and constructing regular expressions (Kleene 1956), to a fairly general algebraic setting (Zimmermann 1981).

The algebraic structure presented in Section 5.2 for cutset enumeration is due to Martelli (1974). Algebraic structures relevant for determining other types of separators in a network are discussed by Carré (1979). The most efficient cutset-enumeration method for undirected networks is that of Tsukiyama *et al.* (1980). In the special case of planar networks, additional computational refinements are available to increase the efficiency of path- and cutset-enumeration techniques (Whited *et al.* 1990).

Several methods have been proposed (Heidtmann 1983; Locks 1978; Rai and Aggarwal 1980) for obtaining the cutsets of a network from a description of its paths, or conversely. In fact, these techniques are applicable to any coherent system, enabling the 'mincuts' of the system to be derived from its 'minpaths' (for a discussion of these latter concepts see Chapter 6). Improved algorithms to carry out this path (or cutset) 'inversion' have been reported by Benzaken (1966) and Shier and Whited (1985).

6

Pseudopolynomial algorithms for calculating reliability

As discussed in previous chapters, the calculation of reliability is known to be #P-complete, meaning that there is unlikely to exist an algorithm that is polynomial in the size of the input network (the number of nodes and the number of edges). Part of the difficulty inherent in reliability calculations is the fact that the number of simple paths and cutsets in a network can grow exponentially with its size. Thus the techniques of inclusion–exclusion and disjoint products, as well as other methods that process a given listing of the paths or cutsets, are automatically doomed to exponential growth in the worst case. A less ambitious, but still important, objective is to seek *pseudopolynomial* algorithms for calculating network reliability—methods that are polynomially bounded in the number of simple paths or the number of cutsets. Since there exist methods, described in Chapter 5, that allow the paths and cutsets to be generated efficiently in terms of the number of such objects, subsequent processing of these objects by a pseudopolynomial algorithm might be effective in cases where the number of such objects is not too large.

The overall aim of this chapter is to investigate algebraic structures that govern the paths and cutsets of a network. If such algebraic structures obey certain properties, then a reliability algorithm can be devised whose complexity is polynomially dependent on the number of these objects. While the two-terminal reliability problem provides a motivating instance for studying such pseudopolynomial algorithms, our approach extends to more general reliability systems, called 'coherent' systems. Section 6.1 introduces the concept of an underlying semilattice through several examples. This algebraic structure permits the development of a pseudopolynomial algorithm, as described in Section 6.2. A new perspective on reliability calculations, based on Möbius inversion applied to the semilattice, is discussed in the final section.

6.1 Examples of lattice structures

To motivate the subsequent development, we first examine the two-terminal network reliability problem and identify some natural lattice structures that can be imposed on the s-t paths and s-t cutsets of the network. For concreteness, consider the undirected 'bridge' network G shown in Fig. 6.1, having source node s and destination node t. As

Examples of lattice structures

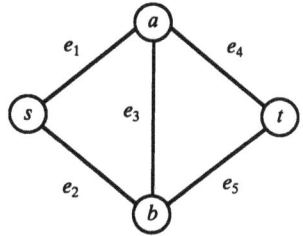

Fig. 6.1 Undirected bridge network.

discussed in the previous chapter, each s-t cutset S in $G = (N, E)$ can be represented as a cut $S = \langle X, \bar{X} \rangle$ in which the (induced) subnetworks on X and on $\bar{X}$ are both connected. That is, the cutset S consists of all edges in E joining nodes of X with nodes of $N - X$. For the present example, there are four s-t cutsets, defined by the following edge/node sets:

$$S_1 = \{e_1, e_2\} \qquad X_1 = \{s\}$$
$$S_2 = \{e_1, e_3, e_5\} \qquad X_2 = \{s, b\}$$
$$S_3 = \{e_2, e_3, e_4\} \qquad X_3 = \{s, a\}$$
$$S_4 = \{e_4, e_5\} \qquad X_4 = \{s, a, b\}.$$

A partial order $\geqslant$ on the edge sets S_i can then be defined by the usual subset relation on the associated node sets: namely

$$S_i \geqslant S_j \Leftrightarrow X_i \supseteq X_j. \tag{6.1}$$

The *Hasse diagram* for this partial order is shown in Fig. 6.2, in which S_1 is the minimal element, S_4 is the maximal element, and relations implied by transitivity are not explicitly shown. As a matter of fact, this partially ordered set forms a *lattice*: any two S_i and S_j have a greatest lower bound $S_i \wedge S_j$ and a least upper bound $S_i \vee S_j$.

Alternatively, the s-t paths of the network in Fig. 6.1 can be partially

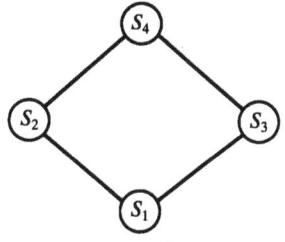

Fig. 6.2 Hasse diagram for the undirected bridge network.

ordered. In this case, the four s-t paths

$$S_1 = \{e_2, e_5\}, \qquad S_2 = \{e_1, e_3, e_5\}, \qquad S_3 = \{e_2, e_3, e_4\}, \qquad S_4 = \{e_1, e_4\}$$

can be partially ordered by their geometric position:

$$S_i \geqslant S_j \Leftrightarrow \text{path } S_i \text{ lies 'above' path } S_j. \tag{6.2}$$

The Hasse diagram for this partial order on paths turns out to be the same lattice already displayed in Fig. 6.2.

As the above examples suggest, we can define a partial ordering on the s-t cutsets in an arbitrary network G using (6.1). In order to make geometric sense of the notion of one path lying 'above' another, we restrict attention to *planar* networks (drawn in the plane without edges crossing) in which s and t lie on the boundary of the exterior region. In such an *s-t planar network*, the s-t paths can be ordered using (6.2). In both cases, the partial order is in fact a lattice.

The fundamental role played by paths and cutsets in network reliability can be viewed more clearly using the general concept of a *coherent binary system*. A coherent (binary) system is defined with respect to some ground set $E = \{e_1, \ldots, e_n\}$ of *components*. Each component e_k assumes one of two states, operational or failed, with probabilities p_k and $q_k = 1 - p_k$ respectively. A *structure function* $\Phi(S)$ is defined on all subsets $S \subseteq E$: $\Phi(S) = 1$ if the system operates when components in S operate and those in $E - S$ fail, and $\Phi(S) = 0$ otherwise. Moreover, the structure function Φ of a coherent system is assumed to satisfy:

$$S \subseteq T \Rightarrow \Phi(S) \leq \Phi(T) \tag{6.3}$$

$$\Phi(\emptyset) = 0, \qquad \Phi(E) = 1. \tag{6.4}$$

The requirement of monotonicity (6.3) simply means that an operating system cannot fail if one of its failed components is repaired (changed to operational). The requirement (6.4) serves to exclude certain trivial special cases: $\Phi(S) \equiv 0$ or $\Phi(S) \equiv 1$. In a coherent system a *minpath* is defined as a minimal set $P \subseteq E$ such that $\Phi(P) = 1$; that is, $\Phi(P) = 1$ but $\Phi(S) = 0$ holds for every proper subset S of P. A *mincut* is a minimal set $K \subseteq E$ such that $\Phi(E - K) = 0$; that is, $\Phi(E - K) = 0$ but $\Phi(E - S) = 1$ holds for every proper subset S of K. Notice that the minpaths of a two-terminal network are just the simple s-t paths, while the mincuts are the s-t cutsets. Because the system is coherent, knowledge of either the minpaths or the mincuts is sufficient to describe the operation of the system. For example, in terms of the minpaths, $\Phi(S) = 1$ precisely when $S \supseteq P$ for some minpath P. The fundamental problem for a coherent system is to calculate its *system reliability* $R_\Phi = \Pr[\Phi(S) = 1]$.

One example of a coherent system is provided by a system with components ordered $e_1, \ldots, e_n$. The system is considered to function whenever some three or more 'consecutive' components operate. Thus the minpaths are given by the following subsets:

$$S_1 = \{e_1, e_2, e_3\}, \quad S_2 = \{e_2, e_3, e_4\}, \quad \ldots, \quad S_{n-2} = \{e_{n-2}, e_{n-1}, e_n\}.$$

In this instance, it is natural to order the subsets linearly: $S_1 \leqslant S_2 \leqslant \cdots \leqslant S_{n-2}$. Clearly this 'chain' forms a lattice, and it does not derive from any network. That is, the sets S_i do not correspond to the paths or cutsets of some network.

In the three examples presented in this section, the cutsets of a network, the paths of an s-t planar network, and the minpaths of a 'consecutive k-out-of-n' system, it has been possible to define a lattice ordering for certain subsets of the component set E. If further properties are assumed for this lattice then reliability can be calculated in pseudo-polynomial time. Certainly some conditions must be imposed on the ordering, since we could, for example, arbitrarily order the paths of any network G. Recall from Section 2.5, however, that there is unlikely to exist any general algorithm for calculating $R_{st}(G)$ that is pseudo-polynomial in the number of paths. The following section discusses additional requirements placed on the lattice and their implications for efficient reliability computations.

6.2 A recursive algorithm for semilattices

Suppose that $E = \{e_1, \ldots, e_n\}$ is a set of components, each of which is subject to random and independent failure. Let $\mathcal{S} = \{S_1, \ldots, S_r\}$ be a specified collection of subsets of E. Rather than requiring these sets to be ordered by a lattice, it is sufficient to suppose only that $(\mathcal{S}, \leqslant)$ forms a *meet semilattice*. Namely, any two S_i, $S_j \in \mathcal{S}$ have a (unique) greatest lower bound $S_i \wedge S_j$ in the partial order $\leqslant$. There are two additional axioms that are imposed here.

$$S_i \leqslant S_k \leqslant S_j, \quad e \in S_i, \quad e \in S_j \implies e \in S_k. \tag{6.5}$$

$$S_i \wedge S_j \subseteq S_i \cup S_j. \tag{6.6}$$

Axiom (6.5) can be rephrased as saying that the collection of sets S_i containing a given component e is *convex*. Axiom (6.6) is a type of closure requirement. To illustrate these conditions, suppose S_i and S_j are two s-t paths in an s-t planar network. Relative to the geometric ordering of paths (6.2), $S_i \wedge S_j$ corresponds to the lower envelope path defined by S_i and S_j (see Fig. 6.3). Requirement (6.6) can then be interpreted as simply saying that the edges of this lower envelope path must derive from either S_i or S_j. Requirement (6.5) means that if an edge e belongs to

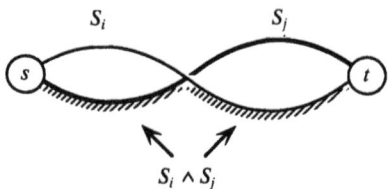

Fig. 6.3 Two s-t paths and their lower envelope path.

paths S_i and S_j, it must also belong to any path that lies between them in the order (6.2). Both these axioms clearly hold then for the path semilattice. It can be easily verified that the axioms hold as well for the cutset semilattice defined by (6.1). In the case of a consecutive k-out-of-n system, whose minpaths are linearly ordered, (6.6) trivially holds since $S_i \wedge S_j = S_{\min(i,j)}$. Moreover, (6.5) holds because any component e appears only in consecutive minpaths S_i.

Suppose now that components in E have two states, called *active* and *inactive*. Let p_k denote the probability that component e_k is active. A set $S \subseteq E$ is called *active* if all its components are active. In particular, for each $S_j \in \mathcal{S}$, the event $\{S_j \text{ is active}\}$ is denoted by E_j. We will be interested in calculating the quantity $p(E, \mathcal{S}) = \Pr(E_1 \cup E_2 \cup \cdots \cup E_r)$, the probability that at least one of the r given subsets S_j is active. In the case of the cutset semilattice, the association of 'active' with 'failing' allows E_j to be interpreted as the event that all edges of cutset S_j fail, whereupon $p(E, \mathcal{S})$ is just the two-terminal unreliability $U_{st}(G)$ of the underlying network G. For the path semilattice, 'active' should be interpreted as 'operating' and then $p(E, \mathcal{S})$ is just the two-terminal reliability $R_{st}(G)$.

As noted in earlier chapters, a fundamental difficulty in calculating $p(E, \mathcal{S})$ is that the events E_j are not, in general, disjoint. However, we can define events F_j that are disjoint by

$$F_j = \{S_j \text{ is the 'lowest' active set in } \mathcal{S}\}.$$

These new events are, in fact, well defined. Namely, if $S_{i_1}, S_{i_2}, \ldots, S_{i_k}$ are active sets then by axiom (6.6) so is $S_{i_1} \wedge S_{i_2} \wedge \cdots \wedge S_{i_k}$, and thus $S_{i_1} \wedge S_{i_2} \wedge \cdots \wedge S_{i_k} = S_{i_j}$, for some j. Notice that $S_{i_1} \wedge S_{i_2} \wedge \cdots \wedge S_{i_k} \leqslant S_{i_l}$ for all l, and so $S_{i_j} \leqslant S_{i_l}$ holds for all active sets S_{i_l}. By the antisymmetry property of $\leqslant$, S_{i_j} is the unique lowest active set in $\mathcal{S}$ relative to the partial order $\leqslant$.

It is now easy to establish the following two properties of the events F_j.

Property 6.1. $F_i \cap F_j = \emptyset$ for $i \neq j$.

A recursive algorithm for semilattices

Proof. If F_i and F_j both occur for $i \neq j$ then S_i and S_j are active, so $S_i \wedge S_j$ is active, by (6.6). Since $S_i \wedge S_j \leq S_i$ and $S_i \wedge S_j \leq S_j$ then $S_i = S_i \wedge S_j = S_j$, a contradiction. □

Property 6.2. $E_1 \cup E_2 \cup \cdots \cup E_r = F_1 \cup F_2 \cup \cdots \cup F_r$.

Proof. Suppose $A = E_1 \cup E_2 \cup \cdots \cup E_r$ occurs with $S_{i_1}, \ldots, S_{i_k}$ being active. Then $S_{i_1} \wedge \cdots \wedge S_{i_k} = S_{i_j}$ is active by (6.6) and so F_{i_j} occurs. Conversely, if F_j occurs then S_j is active so A occurs. □

As a result of the above properties, the events F_j form a partition of the desired event $E_1 \cup E_2 \cup \cdots \cup E_r$ and

$$p(E, \mathscr{S}) = \sum_{j=1}^{r} \Pr(F_j). \tag{6.7}$$

The reliability problem for the semilattice then reduces to that of calculating the values $\Pr(F_j)$. A recursion involving these quantities will now be derived. In what follows, it is assumed without loss of generality that the elements of the partial order have been topologically numbered: that is, $S_i < S_j \Rightarrow i < j$. Also, we define for any $S \subseteq E$

$$\alpha(S) = \prod_{e_k \in S} p_k, \qquad \alpha(\varnothing) \equiv 1. \tag{6.8}$$

Because different components function independently, $\alpha(S)$ is just the probability that S is active. Then we have

$$\begin{aligned}
E_j &= \bigcup_{i=1}^{r} (F_i \cap E_j) \\
&= \left[\bigcup_{S_i < S_j} (F_i \cap E_j) \right] \cup F_j \\
&= \left[\bigcup_{S_i < S_j} F_i \cap \{S_j - S_i \text{ is active}\} \right] \cup F_j. \tag{6.9}
\end{aligned}$$

The first equality above follows from Property 6.2. The second follows since, if $F_i \cap E_j$ occurs then $S_i \wedge S_j = S_i$, whence $S_i \leq S_j$. Because all unions in (6.9) involve disjoint events,

$$\Pr(E_j) = \left[\sum_{S_i < S_j} \Pr(F_i) \Pr(S_j - S_i \text{ is active} \mid F_i) \right] + \Pr(F_j).$$

Now, by requirement (6.5), $e \in S_j - S_i$ cannot be an element of any $S_k < S_i$. This means that the event F_i, which requires S_i to be active and no $S_k < S_i$ to be active, is independent of the event $\{S_j - S_i \text{ is active}\}$,

whence
$$\Pr(E_j) = \left[\sum_{S_i < S_j} \Pr(F_i)\alpha(S_j - S_i)\right] + \Pr(F_j). \tag{6.10}$$

Rearranging (6.10) gives the recursion:
$$\Pr(F_j) = \Pr(E_j) - \left[\sum_{S_i < S_j} \Pr(F_i)\alpha(S_j - S_i)\right], \quad j = 1, \ldots, r. \tag{6.11}$$

This recursion was first suggested by Provan and Ball (1984), and it was originally stated in the specific context of s-t cutsets in a network. The quantities $\Pr(E_j)$ and $\alpha(S_j - S_i)$ are readily computed and thus $\Pr(F_1), \Pr(F_2), \ldots, \Pr(F_r)$ can be found in turn, using (6.11). The quantity $p(E, \mathcal{S})$ is then determined via (6.7). To assess the worst-case complexity of the procedure, suppose that $|E| = n$ and $|\mathcal{S}| = r$. There are at most $O(r^2)$ products to be computed in (6.11), and there is $O(n)$ work to calculate each $\alpha(S_j - S_i)$, giving a worst-case complexity of $O(nr^2)$. The above procedure is thus pseudopolynomial: its running time is bounded by a polynomial in the number of elements in the partial order.

To illustrate the recursive approach, consider the two-terminal network given in the upper portion of Fig. 6.4. The ordering of s-t paths from 'top' to 'bottom' produces the Hasse diagram shown in the lower portion of the figure. The edges defining each path are indicated beside each of the seven s-t paths S_j. For simplicity, we consider the case in which all edges have the same reliability p. By employing the shorthand notation $a_j \equiv \Pr(E_j)$ and $f_j \equiv \Pr(F_j)$, relation (6.11) yields:

$$f_1 = a_1 = p^2,$$
$$f_2 = a_2 - f_1(p^2) = p^3 - p^4,$$
$$f_3 = a_3 - f_1(p^2) = p^3 - p^4,$$
$$f_4 = a_4 - f_1(p^3) - f_2(p^2) = p^4 - 2p^5 + p^6,$$
$$f_5 = a_5 - f_1(p^4) - f_2(p^2) - f_3(p^2) = p^4 - 2p^5 + p^6,$$
$$f_6 = a_6 - f_1(p^3) - f_3(p^2) = p^4 - 2p^5 + p^6,$$
$$f_7 = a_7 - f_1(p^3) - f_2(p^2) - f_3(p^2) - f_4(p) - f_5(p) - f_6(p)$$
$$= p^3 - 6p^5 + 8p^6 - 3p^7,$$

and so $R_{st}(G) = p(E, \mathcal{S}) = \sum_{j=1}^{7} f_j = p^2 + 3p^3 + p^4 - 12p^5 + 11p^6 - 3p^7$.

While the emphasis here has been on *exact* calculation of reliability, the recursive algorithm can also be used to provide efficiently computable

A recursive algorithm for semilattices

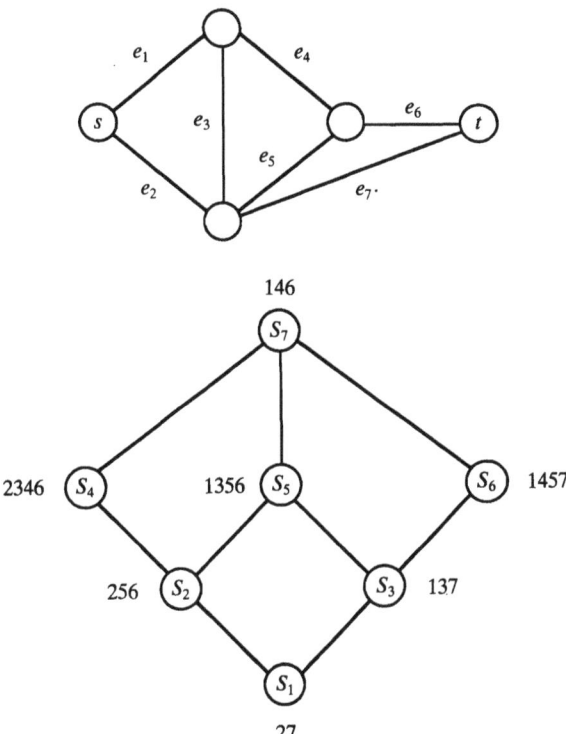

Fig. 6.4 An illustrative network and the Hasse diagram for its s-t paths.

bounds on $R_{st}(G)$. For example, rather than enumerating all s-t cutsets of G (with $s = 1$, $t = n$), we can concentrate on the $n-1$ cuts $S_i = \langle X_i, \bar{X}_i \rangle$ defined by $X_i = \{1, \ldots, i\}$, $1 \leq i < n$. Clearly, these cuts are linearly ordered by (6.1), giving $S_1 \leq S_2 \leq \cdots \leq S_{n-1}$. Each S_i is either an s-t cutset or contains an s-t cutset, so that calculation of $\Pr(E_1 \cup E_2 \cup \cdots \cup E_{n-1})$, where E_i indicates the event that all edges of S_i fail, provides an upper bound on two-terminal reliability. The recursive algorithm, applied to the sublattice defined by $\{S_1, S_2, \ldots, S_{n-1}\}$, can be carried out in $O(nr^2) = O(n^3)$ time, which is now polynomial in the size of the network. As a matter of fact, by taking advantage of the special structure of this linear ordering, it is possible to streamline the recursive algorithm and obtain instead an $O(n^2)$ complexity (Shanthikumar 1988).

To describe this improved algorithm, we will assume, without loss of generality, that every edge $k = (i, j)$, $i < j$, is present in the network and fails with probability $p_k = p_{ij}$. (Nonexistent edges are easily represented by using a failure probability of 1.) Since S_i is the set of all edges joining nodes in $\{1, \ldots, i\}$ to nodes in $\{i+1, \ldots, n\}$, it follows that for $i < j$

$$\alpha_{ij} = \alpha(S_j - S_i) = \prod \{p_{ab} : i < a \leq j, j < b \leq n\}.$$

By convention $\alpha_{jj} = \alpha(\emptyset) = 1$, and so the following relation holds for $i = 2, \ldots, j$:

$$\alpha_{i-1,j} = \alpha_{ij} \beta_{ij}, \qquad (6.12)$$

where β_{ij}, $i \leq j$, is defined by

$$\beta_{ij} = \prod \{p_{ib} : j < b \leq n\},$$

with $\beta_{in} \equiv 1$. Now the β_{ij} can be computed in $O(n^2)$ operations by means of the recursion:

$$\beta_{i,j-1} = \beta_{ij} p_{ij}, \qquad j = n, \ldots, 2; \qquad i = 1, \ldots, j-1.$$

Given the β_{ij}, we can compute the α_{ij} in $O(n^2)$ operations by means of the recursion (6.12), using $j = 2, \ldots, n-1$ and $i = j, \ldots, 2$. Thus the required α_{ij} can be found in $O(n^2)$ time, and solving the lattice equations (6.11) then has $O(n^2)$ overall complexity.

6.3 Möbius inversion and reliability calculations

In this section we will gain an additional perspective on the nature of reliability calculations by applying the technique of Möbius inversion to the partially ordered set $(\mathscr{S}, \leq)$. Before proceeding, recall that a *chain* $C = [S_{i_0}, S_{i_1}, \ldots, S_{i_k}]$ in a partially ordered set is a linearly ordered set of elements: $S_{i_0} \leq S_{i_1} \leq \cdots \leq S_{i_k}$. The *length* $|C|$ of a chain C on $k+1$ elements is defined to be k.

By using the shorthand notation introduced at the end of the last section, eqn (6.11) becomes

$$f_j = a_j - \sum_{S_i < S_j} f_i \alpha_{ij}, \qquad j = 1, \ldots, r,$$

where $\alpha_{ij} = \alpha(S_j - S_i)$. In view of the fact that $\alpha_{jj} = \alpha(\emptyset) = 1$, this can be rearranged as

$$a_j = \sum_{S_i \leq S_j} f_i \alpha_{ij}, \qquad j = 1, \ldots, r. \qquad (6.13)$$

Our objective is to 'invert' this expression, thereby giving a formula for f_j in terms of the a_i:

$$f_j = \sum_{S_i \leq S_j} a_i \mu_{ij}, \qquad j = 1, \ldots, r. \qquad (6.14)$$

An approach for carrying out this inversion over a partially ordered set is called Möbius inversion (Rota 1964). We briefly indicate how this is accomplished and provide, as well, an explicit interpretation for the coefficients μ_{ij}.

First, it is convenient to define $\alpha_{ij} = 0$ whenever $S_i \leqslant S_j$ does not hold. Thus (6.13) describes a system of r equations relating the row vector $\mathbf{a} = (a_1, a_2, \ldots, a_r)$ to the row vector $\mathbf{f} = (f_1, f_2, \ldots, f_r)$ via $\mathbf{a} = \mathbf{f}A$, where $A = (\alpha_{ij})$. Since the elements of the partial order are assumed to be topologically numbered, the coefficient matrix A is unit upper triangular, so that $A = I + U$, where I is the identity matrix and U is a strictly upper triangular matrix. Now A is nonsingular and U is nilpotent of index at most r, so we can express

$$\mathbf{f} = \mathbf{a}A^{-1} = \mathbf{a}[I + U]^{-1} = \mathbf{a}[I - U + U^2 + \cdots + (-1)^{r-1}U^{r-1}]. \quad (6.15)$$

As can be seen in the proof of Theorem 3.4, the i-j entry of U^k is simply the sum of products $\pi(C) = \alpha_{i,i_1}\alpha_{i_1,i_2}\cdots\alpha_{i_{k-1},j}$ over all chains $C = [S_i, S_{i_1}, \ldots, S_{i_{k-1}}, S_j]$ of length k extending between S_i and S_j. Comparison of (6.14) with (6.15) shows that μ_{ij} is the alternating sum of such chain products, taken over all chains C in $(\mathcal{S}, \leqslant)$ joining S_i and S_j:

$$\mu_{ij} = \begin{cases} 1, & \text{if } i = j \\ \sum_C (-1)^{|C|}\pi(C), & \text{if } i \neq j. \end{cases} \quad (6.16)$$

From $\alpha_{ij} = \alpha(S_j - S_i)$ and axiom (6.5), it is seen that the product $a_i\pi(C)$ for the chain $C = [S_i, S_{i_1}, \ldots, S_{i_{k-1}}, S_j]$ simplifies to

$$\Pr(S_i \text{ active}) \Pr(S_{i_1} - S_i \text{ active}) \cdots \Pr(S_j - S_{i_{k-1}} \text{ active})$$
$$= \Pr(S_i \cup S_{i_1} \cup \cdots \cup S_j \text{ active}).$$

Using the notation $\alpha(C)$ for the 'telescoped' product on the right, we then have an equivalent expression for (6.14):

$$f_j = \sum_C (-1)^{|C|}\alpha(C), \quad (6.17)$$

where the summation is over all chains C extending downward from S_j in $(\mathcal{S}, \leqslant)$. Consequently, from (6.7)

$$p(E, \mathcal{S}) = \sum_{j=1}^r f_j = \sum_C (-1)^{|C|}\alpha(C), \quad (6.18)$$

where C now ranges over chains of all lengths in $(\mathcal{S}, \leqslant)$.

Equation (6.18) thus provides an interpretation of system reliability (or unreliability) as an alternating sum of appropriate product terms over all chains in the underlying semilattice. To illustrate this expansion, consider the directed bridge network shown in Fig. 6.5, whose associated path

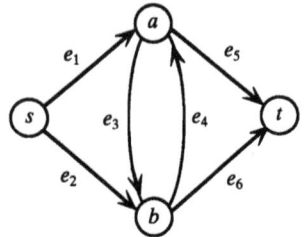

Fig. 6.5 The directed bridge network.

semilattice is given by the Hasse diagram in Fig. 6.6, relative to the four s-t paths

$$S_1 = \{e_2, e_6\}, \quad S_2 = \{e_1, e_3, e_6\}, \quad S_3 = \{e_2, e_4, e_5\}, \quad S_4 = \{e_1, e_5\}.$$

Application of (6.17) yields, for example,

$$f_4 = p_1 p_5 - (p_1 p_3 p_5 p_6 + p_1 p_2 p_4 p_5 + p_1 p_2 p_5 p_6) + p_1 p_2 p_3 p_5 p_6 + p_1 p_2 p_4 p_5 p_6,$$

where the first term corresponds to the chain of length 0 at S_4, the next (negated) terms correspond to the three chains of length 1 extending down from S_4, and the final two terms correspond to the chains of length 2 from S_4. The expansion for the two-terminal reliability $p(E, \mathcal{S})$ obtained by applying (6.18) has a total of 11 terms, corresponding to the 11 chains in $(\mathcal{S}, \leqslant)$. This number is to be compared with the $2^4 - 1 = 15$ terms potentially appearing in the inclusion–exclusion formula for $\Pr(E_1 \cup \cdots \cup E_4)$. Thus eqn (6.18) captures some of the cancellation found in the topological expansion formula expressed by Theorem 2.1.

As another application of eqn (6.18), consider any directed network with s-t cutsets given by $S_1, \ldots, S_r$. Recall from Section 6.1 that the cutsets $S_i = \langle X_i, \bar{X}_i \rangle$ can be ordered by set inclusion with respect to their

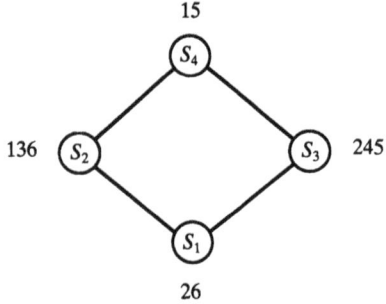

Fig. 6.6 The Hasse diagram for s-t paths in the directed bridge network.

associated node sets X_i. Then the inversion formula (6.18) expresses the system unreliability as an alternating sum of terms, each corresponding to a sequence of cutsets $S_{i_0}, S_{i_1}, \ldots, S_{i_k}$ having node sets satisfying $X_{i_0} \subset X_{i_1} \subset \cdots \subset X_{i_k}$, $s \in X_{i_0}$, $t \notin X_{i_k}$. This node-based expansion of the inclusion–exclusion formula for cutsets has been investigated by Buzacott (1987).

6.4 Chapter notes

This chapter has studied the lattice structure of paths and cutsets in networks, focusing on the important properties of these lattices that lead to relatively efficient algorithms for computing reliability. Namely, the worst-case complexity of such algorithms grows only polynomially with the number of such objects. Buzacott (1983) initially noticed that the disjoint-products formula could be simplified if the cutsets were first partially ordered and then processed in this order. Shanthikumar (1988) also employed this partial ordering on cutsets to approximate the two-terminal reliability of a network. A good description of general coherent systems, and properties of its minpaths and mincuts, can be found in Barlow and Proschan (1981) and also in Ross (1985).

Provan and Ball (1984) were the first to develop a recursive algorithm, running in pseudopolynomial time, based on the structure of s-t cutsets in an arbitrary network. This work was further generalized to a lattice setting by Shier (1988a), in which computational results with the recursive approach are also reported. Ball and Provan (1987) also generalized this work to obtain a pseudopolynomial algorithm for computing K-terminal reliability in directed or undirected networks. For further discussion of partially ordered sets and Möbius inversion, consult the paper of Greene (1982) and Chapter 3 of Stanley (1986).

7

Reliability covering problems

This chapter studies a type of reliability problem in which a given set S needs to be covered by certain of its subsets. The subsets are available for use on a random and independent basis with known probabilities. It is of interest to calculate the probability that the entire set S is 'covered' by the operating subsets: i.e., every element of S is present in some available subset. As an example of such a problem, consider a mass transit system containing a number of bus routes and the stops they serve. Because of maintenance or staffing problems, a particular bus route might not always be in service: specifically the route is only known to operate (or be available) with some fixed probability. A useful performance measure for the transit system is therefore the probability that each stop is served by some operating bus route. There are a number of other possible applications of this model, such as evaluating the reliability of delivery routes (needed to cover customer demands) or determining the reliability of flight schedules for aircraft (serving airports). Another application is discussed more fully in Section 7.1, and a general description of the reliability covering problem is presented in Section 7.2. It is shown that solving the covering problem is, in fact, equivalent to calculating the reliability of a coherent system, which is known to be mathematically difficult. Consequently, the general covering problem is provably hard. As demonstrated in Section 7.2, the problem remains hard even when defined on fairly specialized systems (tree networks). One solvable case, which derives from the semilattice approach developed in the previous chapter, is discussed in Section 7.3, while another solvable case is pursued in Section 7.4.

7.1 A motivating example

Suppose that an interval I on the real line is given, together with a collection of subintervals $I_1, I_2, \ldots, I_n$ of this interval. In addition, suppose that each subinterval I_k is randomly available for use with probability p_k. Then what is the probability that all available subintervals cover the entire interval I? This type of problem might arise, for instance, in maintaining continuous surveillance of a critical portion of a country's border. Located near the border are several guard posts, each of which can monitor some segment of the border. In this case, the

A motivating example

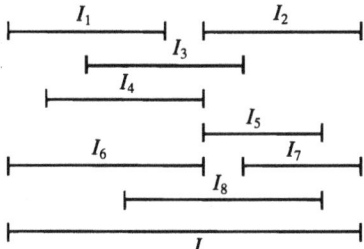

Fig. 7.1 Covering an interval I by certain subintervals I_k.

interval I corresponds to the critical portion of the border and each subinterval I_k corresponds to the (contiguous) segment of the border monitored by post k. The requirement of continuous surveillance is simply the condition that all operating posts collectively cover the critical border segment.

Fig. 7.1 shows a specific example involving eight subintervals which will be used throughout the chapter to illustrate several solution approaches. This subinterval problem can equivalently be phrased as a covering problem on an associated network G. Namely, the endpoints of all subintervals I_k are first projected onto I, and a node is then identified with each resulting subdivision of I (see Fig. 7.2). As a result, each of the original subintervals I_k corresponds to a subpath P_k spanning certain nodes of the path network G. The original subinterval covering problem is then equivalent to the problem of stochastically covering all nodes of G using the eight subpaths P_k, each of which is randomly available with probability p_k.

A possible solution strategy for this network covering problem is based on considering each node of G in turn. In order for node a to be covered, either subpath 1 or subpath 6 must operate; in order for node b to be covered, one of the subpaths 1, 4, or 6 must operate; and so on. Suppose

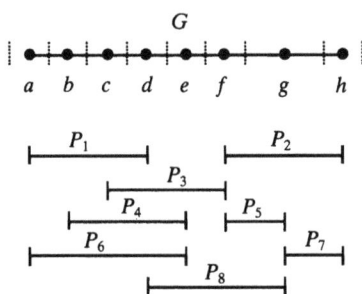

Fig. 7.2 An equivalent network covering problem.

that the symbol j is used to denote the event that subpath j operates and the symbol $\bar{j}$ is used to denote the event that subpath j does not. The compound event E that *all* nodes of G are covered is then given by

$$E = (1 \cup 6)(1 \cup 4 \cup 6)(1 \cup 3 \cup 4 \cup 6) \cdots (2 \cup 5 \cup 7 \cup 8)(2 \cup 7)$$
$$= (1 \cup 6)(3 \cup 4 \cup 6 \cup 8)(2 \cup 3 \cup 5 \cup 8)(2 \cup 7), \tag{7.1}$$

using the standard rules for simplifying Boolean expressions. Further application of the properties of Boolean algebras yields

$$E = (6 \cup 13 \cup 14 \cup 18)(2 \cup 37 \cup 57 \cup 78)$$
$$= 26 \cup 123 \cup 124 \cup 128 \cup 367 \cup 137 \cup 567 \cup 1457 \cup 678 \cup 178.$$

The required $\Pr(E)$ could then be found by applying the inclusion–exclusion formula, but it is clear that a great deal of manipulation would be needed to process the $2^{10} - 1$ terms of the expansion.

Alternatively, we could apply the disjoint-products technique to evaluate $\Pr(E)$. It is convenient to work instead with the complementary event $\bar{E}$, given by applying DeMorgan's laws to (7.1):

$$\bar{E} = \bar{1}\bar{6} \cup \bar{3}\bar{4}\bar{6}\bar{8} \cup \bar{2}\bar{3}\bar{5}\bar{8} \cup \bar{2}\bar{7} = E_1 \cup E_2 \cup E_3 \cup E_4, \tag{7.2}$$

so that

$$\Pr(\bar{E}) = \Pr(E_1) + \Pr(\bar{E}_1 E_2) + \Pr(\bar{E}_1 \bar{E}_2 E_3) + \Pr(\bar{E}_1 \bar{E}_2 \bar{E}_3 E_4).$$

$\Pr(\bar{E})$, and consequently $\Pr(E) = 1 - \Pr(\bar{E})$, can be found by evaluating these four probabilities in terms of the failure probabilities $q_k = 1 - p_k$.

$$\Pr(E_1) = \Pr(\bar{1}\bar{6}) = q_1 q_6,$$
$$\Pr(\bar{E}_1 E_2) = \Pr([1 \cup 6]\bar{3}\bar{4}\bar{6}\bar{8}) = \Pr(1\bar{3}\bar{4}\bar{6}\bar{8}) = p_1 q_3 q_4 q_6 q_8,$$
$$\Pr(\bar{E}_1 \bar{E}_2 E_3) = \Pr([1 \cup 6][3 \cup 4 \cup 6 \cup 8]\bar{2}\bar{3}\bar{5}\bar{8}) = \Pr([1 \cup 6][4 \cup 6]\bar{2}\bar{3}\bar{5}\bar{8})$$
$$= \Pr([6 \cup 14]\bar{2}\bar{3}\bar{5}\bar{8}) = \Pr([6 \cup 14\bar{6}]\bar{2}\bar{3}\bar{5}\bar{8})$$
$$= q_2 q_3 q_5 p_6 q_8 + p_1 q_2 q_3 p_4 q_5 q_6 q_8,$$

where the third probability requires a second invocation of the disjoint-products formula.

$$\Pr(\bar{E}_1 \bar{E}_2 \bar{E}_3 E_4) = \Pr([1 \cup 6][3 \cup 4 \cup 6 \cup 8][2 \cup 3 \cup 5 \cup 8]\bar{2}\bar{7})$$
$$= \Pr([1 \cup 6][3 \cup 8 \cup 45 \cup 56]\bar{2}\bar{7})$$
$$= \Pr([13 \cup 18 \cup 145 \cup 36 \cup 68 \cup 56]\bar{2}\bar{7}).$$

Further processing of this last expression is too burdensome to continue, as it will involve considerable effort in expansion and simplification in addition to repeated applications of the disjoint-products technique.

The point of these incomplete solution attempts has been to demonstrate that direct use of the standard inclusion–exclusion and disjoint-products methods presents a significant computational chore, even for such a simple example as that given in Fig. 7.1. Sections 7.3 and 7.4 show that efficient algorithms can be fashioned to solve such 'linear subpath' problems, as well as certain other covering problems. However, as will be seen in Section 7.2, the general covering problem is equivalent to a more familiar reliability problem that is known to be inherently difficult. Thus there is little hope of developing a polynomial-time algorithm for the general reliability covering problem. The next section formally defines the general covering problem and demonstrates that the problem remains mathematically difficult even under some fairly restrictive conditions.

7.2 The complexity of covering problems

To describe the general *reliability covering problem* (RCP), we use terminology suggested by the mass transit application described earlier. Let $S = \{s_1, s_2, \ldots, s_m\}$ be a set of elements (*stops*), and let $\mathcal{R} = \{R_1, R_2, \ldots, R_n\}$ be a family of subsets of S (*routes*). Each R_k in $\mathcal{R}$ fails randomly and independently with probability q_k. It is desired to calculate the probability $p(S, \mathcal{R})$ that every stop $s_j \in S$ is covered by an operating route:

$$p(S, \mathcal{R}) = \Pr(\cup \{R_k : R_k \text{ is operating}\} = S).$$

This problem is, in fact, equivalent to a reliability problem defined on a coherent binary system. Recall that a coherent system involves a set $C = \{c_1, c_2, \ldots, c_n\}$ of components, each of which fails randomly and independently. A *success set* for the system is any subset $X \subseteq C$ such that the system operates when all components in X operate and all components in $C - X$ fail. Because the system is coherent, it can be represented by $(C, \mathcal{P})$, where $\mathcal{P}$ contains the minimal success sets (minpaths). A *failure set* $Y \subseteq C$ is such that the system fails when all components in Y fail and all components in $C - Y$ operate. Again, by the assumption of coherence, it suffices to know the collection of minimal failure sets, or mincuts, $\mathcal{K} = \{K_1, K_2, \ldots, K_m\}$ to completely describe the given system. The representation of the system as $(C, \mathcal{K})$ provides a 'dual' type of characterization of the success sets: a set X of operating components is a success set if and only if X intersects every element of $\mathcal{K}$.

Given an instance of the RCP, construct a coherent system where component c_k corresponds to route R_k and the mincut K_j corresponds to the routes containing stop s_j. (If the associated K_j is not minimal, then K_j should be discarded.) Under this identification a set X of routes is a cover for $(S, \mathcal{R})$ precisely when $X \cap K_j \neq \emptyset$ for all j. From the dual characterization mentioned above, a set X of components with this property is a success set for the system $(C, \mathcal{K})$. Conversely, each success set of $(C, \mathcal{K})$ yields a route cover of $(S, \mathcal{R})$. Thus calculating the probability $p(S, \mathcal{R})$ that the RCP has a route cover is identical to calculating the reliability of the coherent system $(C, \mathcal{K})$.

It is also true that the reliability problem for a coherent system can be described as a reliability covering problem. Given the coherent system $(C, \mathcal{K})$, identify a stop s_j with each set K_j and a route R_k with each component c_k. The set of stops in R_k will consist of those s_j such that K_j contains c_k. Every success set of (operating) components for $(C, \mathcal{K})$ yields a cover of (operating) routes for $(S, \mathcal{R})$, and conversely. Consequently, the reliability of $(C, \mathcal{K})$ can be determined by solving the associated RCP.

Thus the reliability covering problem is equivalent to the coherent system reliability problem, with the latter expressed in terms of its mincuts. This coherent system reliability problem has been shown (Ball and Provan 1988) to be #P-complete, so that the general reliability covering problem is #P-complete as well. As a result, the remainder of this section will consider special cases of the RCP, in an effort to identify instances that have polynomial complexity. Our overall objective will be to examine the boundary of problem classes separating the polynomially-solvable cases from the #P-complete cases.

In its simplest form, with each R_k being a singleton set, the RCP is easily solved. In this situation $p(S, \mathcal{R}) > 0$ if and only if $\cup R_k = S$, in which case $p(S, \mathcal{R})$ is given by

$$p(S, \mathcal{R}) = \prod_{t=1}^{m} \left[1 - \prod_{s_t \in R_k} q_k \right].$$

By contrast, the reliability covering problem becomes #P-complete when each R_k contains two elements, as will now be demonstrated.

Given a covering system $(S, \mathcal{R})$ with each route having cardinality two, construct an associated undirected network $G = (N, E)$ by identifying node $j \in N$ with stop $s_j \in S$ and identifying edge $[i, j] \in E$ with route $R_k = \{s_i, s_j\} \in \mathcal{R}$. Then a set of routes is a cover of the stops if and only if the corresponding set of edges is an *edge cover* of the nodes of G: a set of edges that is incident with every node of N. We are thus led to study the complexity of the following counting problem.

Number of edge covers (#EC)

INPUT: an undirected network $G = (N, E)$
OUTPUT: the number of edge covers of G
$|\{F \subseteq E: \text{for each } i \in N, \text{ there exists } j \in N \text{ such that } [i, j] \in F\}|.$

Theorem 7.1. *Problem #EC is #P-complete.*

Proof. Given $G = (N, E)$ construct, for each $r = 1, \ldots, |E|$, the network $G^{(r)} = (N, E^{(r)})$, where $E^{(r)}$ contains r parallel copies of each edge in E. Denote the number of edge covers of $G^{(r)}$ by g_r and the number of edge covers of G using exactly j edges by f_j. Since every edge cover of G with j edges produces exactly $(2^r - 1)^j$ edge covers of $G^{(r)}$, the following equations hold:

$$g_r = \sum_{j=1}^{|E|} f_j (2^r - 1)^j, \qquad r = 1, \ldots, |E|.$$

The above linear system has a nonsingular (in fact Vandermonde) coefficient matrix, and we can solve in polynomial time for the f_j given the values g_r. Thus an efficient algorithm for computing g_r would imply an efficient method for calculating (in particular) f_j with $j = |N|/2$, the number of *perfect matchings* in an arbitrary network G. However, this latter counting problem is known to be #P-complete (Valiant 1979), and problem #EC must be #P-complete as well. □

Theorem 7.1 can now be used to show that the RCP, even with each $|R_k| = 2$, is *#P-hard*: i.e., it is at least as hard as any #P-complete problem. To see this, consider the case when all route failure probabilities $q_k = q = 0.5$. Then the probability of a route cover of $(S, \mathcal{R})$, or equivalently the probability of an edge cover of the associated $G = (N, E)$, is given by

$$p(S, \mathcal{R}) = \sum_{j=1}^{|E|} f_j (1-q)^j q^{|E|-j} = 2^{-|E|} \sum_{j=1}^{|E|} f_j,$$

where f_j is the number of edge covers of G with exactly j edges. Then $2^{|E|} p(S, \mathcal{R})$ equals the number of edge covers of G. A polynomial algorithm for calculating $p(S, \mathcal{R})$ would thus yield a polynomial algorithm for counting the number of edge covers of a network, which by Theorem 7.1 is a #P-complete problem. This shows that the RCP with each $|R_k| = 2$ is #P-complete when all failure probabilities are equal. In the general case of arbitrary failure probabilities, the RCP is then #P-hard. By virtue of the isomorphism between the RCP and reliability computations for coherent binary systems. By virtue of the isomorphism between the RCP and reliability computations for coherent binary systems, it follows that the latter calculation is #P-hard even for the special case in which each component (route) appears in exactly two mincuts (stops) or in exactly two minpaths. It is known (Ball and Provan 1988) that such reliability computations are also difficult in the special case when (dually) all mincuts, or all minpaths, have cardinality two.

Problem #EC remains #P-complete when the underlying network G is bipartite, since the related counting problem (number of perfect matchings) remains #P-complete for such networks and the introduction of multiple edges maintains the bipartite character of the networks $G^{(r)}$. Thus the RCP remains difficult even when the routes are defined by the edges of a bipartite network. It might be hoped that by further restricting attention to trees, a special type of bipartite network, a polynomial algorithm might be possible. The following result shows that for undirected trees and routes R_k satisfying $|R_k| = 3$, the problem continues to be #P-hard.

Theorem 7.2. *The RCP is #P-hard on undirected trees, even when routes are paths of cardinality three in the tree.*

Proof. Given $G = (N, E)$, construct the tree T having node set $N \cup \{s\}$ and edges $[s, i]$ for each $i \in N$. Associate with each edge $[i, j] \in E$ the route $R_{ij} = \{i, s, j\}$ in T. Then there is a one-to-one correspondence between edge covers of G and routes covering all nodes of T. As in the previous arguments, the existence of an efficient method to calculate $p(S, \mathcal{R})$ would give an efficient method for counting the number of covering routes in T and thus the number of edge covers in G. The #P-hardness of the RCP for this special case then follows from Theorem 7.1. □

In the proof of Theorem 7.2, the constructed tree T has one node (node s) of extremely high degree. This situation, however, is not the reason for the difficulty of the RCP on trees. Indeed, Theorem 7.2 can be modified to show that if routes consist of paths of arbitrary length in an undirected tree T, the reliability covering problem remains #P-hard, even if all nodes in T have degree at most three.

Finally, we observe that the RCP remains difficult for directed trees T in which routes are directed paths on T. Namely, suppose that $G = (N, E)$ is bipartite with the node set bipartition $N = A \cup B$. Construct the directed tree T with node set $N \cup \{s\}$, and having directed edges (a, s) for $a \in A$, and (s, b) for $b \in B$. For each undirected edge $[a, b] \in E$, with $a \in A$ and $b \in B$, define the (directed) route $a \to s \to b$ in T. A set of such routes will cover all nodes of T if and only if the corresponding edges form an edge cover of G. Then, as in the proof of Theorem 7.2, a polynomial algorithm for counting the number of covering routes in T would solve problem #EC in the bipartite network G. Since the latter problem is #P-complete, the RCP remains #P-hard on directed trees, with routes being directed paths.

Theorem 7.3. *The RCP is #P-hard on directed trees, even when routes are directed paths of cardinality three in the tree.*

Given these negative results, applicable when the routes are paths defined with respect to an underlying tree of stops, it is next appropriate to study what further restrictions might allow an efficient algorithm for the RCP. It turns out that the problem described in Section 7.1, covering a path with certain of its subpaths, can be solved in polynomial time (Hwang and Yao 1989; Shanthikumar 1987). It is possible to generalize this particular case of the RCP in different ways, while still maintaining polynomial complexity. The next two sections develop efficient algorithms for two separate generalizations of reliability covering problems on more inclusive classes of trees than simply paths.

7.3 An algorithm for undirected trees

In this section we study a polynomially-solvable case of the RCP in which routes are defined by certain *subtrees* of a given undirected tree T. This situation includes as a special case the motivating problem studied in Section 7.1, covering a path using certain of its subpaths. In view of Theorem 7.2, which implies that stochastically covering an undirected tree T with arbitrary routes is difficult, some restrictions will need to be placed on the types of routes allowed. Appropriate restrictions will be derived here by applying the semilattice approach presented in Chapter 6. The key idea is to translate the semilattice problem, phrased in terms of calculating coherent system reliability, into the realm of the RCP by invoking the isomorphism between the RCP and the coherent system reliability problem.

To begin, we recall the semilattice framework developed in Chapter 6, applied to the mincuts of a coherent system. Suppose that $(C, \mathcal{K})$ is a coherent system with components $C = \{c_1, c_2, \ldots, c_n\}$ and mincuts $\mathcal{K} = \{K_1, K_2, \ldots, K_m\}$. The collection $\mathcal{K}$ is assumed to be endowed with a partial ordering $\leq$ that forms a meet semilattice. Namely, any two K_i and K_j have a greatest lower bound $K_i \wedge K_j$. If this semilattice also satisfies the following two properties then the $O(nm^2)$ algorithm of Section 6.2 can be applied to calculate the reliability of the system $(C, \mathcal{K})$.

Property I. $K_i \leq K_r \leq K_j$, $c \in K_i$, $c \in K_j \Rightarrow c \in K_r$.

Property II. $K_i \wedge K_j \subseteq K_i \cup K_j$.

We now translate these requirements into the context of the RCP, by identifying route $R_k \in \mathcal{R}$ with component c_k and identifying the routes containing stop $s_j \in S$ with cutset K_j. This means that the stops S of the RCP should be partially ordered by $\leq$, forming a semilattice. Moreover, the above properties can be restated in a fairly appealing manner:

Property I'. $s_i \leq s_r \leq s_j$, $s_i \in R$, $s_j \in R \Rightarrow s_r \in R$.

Property II'. $s_i \notin R$, $s_j \notin R \Rightarrow s_i \wedge s_j \notin R$.

Property I' simply states that each route R is a *convex* set with respect to the ordering $\leq$. Property II' states a *closure* condition: namely, the set $\bar{R} = S - R$ of all stops not in route R forms a sublattice under $\wedge$.

We are interested in examining the implications of these properties in the case when the stops form an undirected *rooted* tree T. That is, the stops correspond to the nodes of a tree semilattice T, with the root r placed at the 'bottom' of the tree. The partial order relation $s_i \leq s_j$ signifies that s_i is on the path joining r and s_j in T. Equivalently, node s_i lies 'below' node s_j in T. Given this implicit orientation, the meet $s_i \wedge s_j$ of two nodes is simply the last common node on the paths joining r to s_i and r to s_j. Fig. 7.3 illustrates a tree semilattice in which the root is s_1 and (for example) $s_6 \wedge s_8 = s_3$. Throughout, a *subtree* of T will mean any connected subnetwork of nodes in T: e.g., in Fig. 7.3 the subnetwork defined by nodes $\{s_3, s_5, s_7, s_9\}$ and all edges of T joining them. The *rooted subtree* T_s indicates the (node) *maximal* subtree of T rooted at some node s: e.g., in Fig. 7.3 the subnetwork defined by nodes $\{s_3, s_5, s_6, s_7, s_8, s_9\}$ and all edges of T joining them. The next two facts concerning tree semilattices can be established without difficulty.

Lemma 7.4. *A subset W of stops is convex if and only if W is the disjoint union of subtrees of T each pair of which have incomparable roots (with respect to $\leq$).*

Lemma 7.5. *A subset W of stops is closed if and only if W is the disjoint union of subtrees of T with each pair of subtree roots r_i and r_j satisfying $r_i \wedge r_j \in W$.*

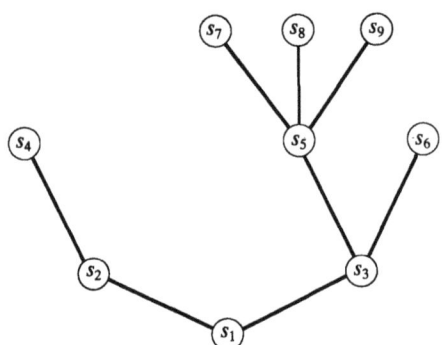

Fig. 7.3 A tree semilattice on nine stops.

Given the above results, we now investigate what route structures $\mathcal{R}$ defined on T satisfy Properties I' and II'. Because $R \in \mathcal{R}$ must be convex, and is thus a union of subtrees with incomparable roots, it suffices to consider the case when R is a single subtree with root r. Note that $R \subseteq T_r \subseteq T$ and $r \in R$. Clearly T_r is closed and (by assumption) so is $\bar{R}$, so that $T_r \cap \bar{R} = T_r - R$ is closed as well. The subnetwork $T_r - R$ in general represents a union of subtrees. We now show that $T_r - R$ actually contains at most one subtree. Suppose, to the contrary, that $T_r - R$ consists of more than one subtree. Then, by the closure of $T_r - R$ and Lemma 7.5, there must be two such subtrees with comparable roots, say r_1 and r_2 with $r_1 \leq r_2$. Because the subtrees are distinct, there exists a node $y \in R$ with $r_1 \leq y \leq r_2$. However, this contradicts the assumed convexity of R since $r \leq r_1 \leq y$ with $r, y \in R$ and $r_1 \notin R$.

Thus $T_r - R$, if nonempty, consists of a single subtree, having some root node $u \in T_r$. We further claim that $T_r - R$ is precisely the rooted subtree T_u of T. Suppose, to the contrary, that some $z \in T_u$ is not in $T_r - R$. Thus $r \leq u \leq z$ with $r, z \in R$ and $u \notin R$, contradicting the convexity of R. Hence $T_r - R$ must (if nonempty) be of the form T_u, where $r \leq u$. Consequently, R can be expressed as $T_r - T_u$ with $r \leq u$. (The possibility of T_u being an empty subtree is also permitted.) Conversely, any route R of this form satisfies Properties I' and II', and we have characterized the tree semilattices that fulfill the desired properties.

Theorem 7.6. *A route R satisfies Properties I' and II' if and only if $R = \bigcup_k [T_{r_k} - T_{u_k}]$, where $r_k \leq u_k$ and the roots $\{r_k\}$ are incomparable in $\leq$.*

If the routes of a system are defined with respect to some tree T and satisfy the requirements of Theorem 7.6, then a polynomial algorithm for the RCP can be fashioned by adapting the algorithm given in Section 6.2. For purposes of exposition, suppose that the routes are labelled $1, \ldots, n$ with q_k being the probability that route k fails. Let X_j denote all routes involving stop s_j, $j = 1, \ldots, m$. Fig. 7.4 displays an illustrative tree

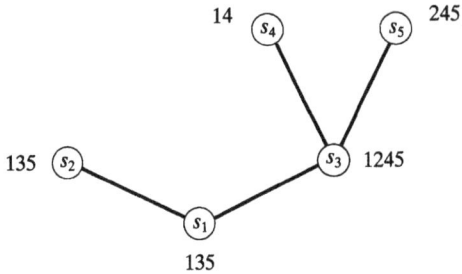

Fig. 7.4 A tree semilattice satisfying Theorem 7.6.

semilattice having five stops and five routes; the sets X_j are indicated beside each stop. It is straightforward to verify that the given routes satisfy the conditions of Theorem 7.6. For example, route 3 can be expressed as $T_{s_1} - T_{s_3}$, the set-theoretic difference between the subtrees rooted at stops s_1 and s_3.

To present the RCP algorithm, we define, for any set of routes X,

$$\alpha(X) = \prod_{k \in X} q_k, \qquad \alpha(\emptyset) = 1.$$

Then the following algorithm correctly determines the coverage probability of an undirected tree T, given that the routes are defined by appropriate subtrees of T.

RCP algorithm I. This algorithm calculates $p(S, \mathcal{R})$ for a rooted undirected tree T with stops $S = \{s_1, s_2, \ldots, s_m\}$ and routes $\mathcal{R} = \{1, 2, \ldots, n\}$ satisfying the conditions of Theorem 7.6. The routes k fail randomly and independently with probability q_k.

1. Place the stops $s_1, s_2, \ldots, s_m$ in topological order so that if $s_i < s_j$ then $i < j$. Let X_j denote the set of routes containing stop s_j.

2. For $j = 1, 2, \ldots, m$

$$f_j = \alpha(X_j) - \sum_{s_i < s_j} f_i \alpha(X_j - X_i).$$

3. Output $p(S, \mathcal{R}) = 1 - \sum_{j=1}^{m} f_j$.

Note that this algorithm begins at the root of T and progressively moves toward the leaves of T. The quantity f_j needed in Step 2 can thus be computed using the (previously determined) values f_i for nodes s_i along the path from the root to node s_j. The validity of this algorithm follows from the development in Section 6.2. As also established there, its worst-case complexity is $O(nm^2)$, which is polynomial in the number of routes n and the number of stops m. The storage requirements for the procedure are $O(m)$.

To illustrate this approach, we solve the subinterval covering problem of Section 7.1. By eqn (7.1), it suffices to consider the RCP defined on the rooted tree shown in Fig. 7.5. Here the routes are certain subpaths of the given path; those routes using stop j are indicated beside each stop. Notice that since the routes are all subpaths of the given path the requirements of Theorem 7.6 are certainly fulfilled. For convenience it is assumed that all routes R_k fail with the same probability $q_k = q$.

An algorithm for directed trees

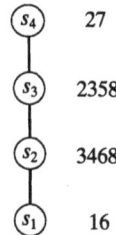

Fig. 7.5 Tree semilattice for the example of Fig. 7.1.

Application of the above RCP algorithm then produces

$$f_1 = \alpha(16) = q^2,$$
$$f_2 = \alpha(3468) - f_1\alpha(348) = q^4 - q^5,$$
$$f_3 = \alpha(2358) - f_1\alpha(2358) - f_2\alpha(25) = q^4 - 2q^6 + q^7,$$
$$f_4 = \alpha(27) - f_1\alpha(27) - f_2\alpha(27) - f_3\alpha(7)$$
$$= q^2 - q^4 - q^5 - q^6 + 3q^7 - q^8,$$
$$p(S, \mathcal{R}) = 1 - \Sigma f_j = 1 - 2q^2 - q^4 + 2q^5 + 3q^6 - 4q^7 + q^8.$$

For instance, if all subintervals in the problem of Fig. 7.1 were randomly available with probability $q = 0.5$, then the coverage probability is computed to be 0.5195.

7.4 An algorithm for directed trees

In this section the RCP is defined on a *rooted directed tree*, T, with root node r. That is, there is exactly one edge of the tree entering every node $i \neq r$ and there are no edges entering node r. Now the routes will correspond to a collection of directed paths in T. Unlike the situation discussed in the previous section, in which routes correspond to fairly broad collections of stops (nearly full subtrees), the present instance allows narrow routes (no branching permitted). A polynomial algorithm will be developed here for computing the probability that all stops (nodes) of the rooted directed tree are covered by operating paths. In the special case when the tree T is a directed path, the routes are then directed subpaths and so the associated RCP again includes the problem of covering a path using a collection of subpaths.

We first establish some necessary terminology. Suppose T forms a rooted directed tree with node set $S = \{1, 2, \ldots, m\}$. A collection $\mathcal{R} = \{R_1, R_2, \ldots, R_n\}$ of directed paths in T is given, and each R_k fails randomly and independently with probability q_k. As defined earlier, T_j denotes the maximal subtree of T rooted at node j. The disjoint union of subtrees obtained by removing the root node j from T_j will be denoted by $T'_j = T_j - \{j\}$. Node y is said to be a *successor* of node x if there is a directed path in T from x to y; y is an *immediate successor* of x if (x, y) is an edge of T. The set of all paths R_k originating at node j will be indicated by $R(j)$, while the set of all paths originating at j or a successor of j will be indicated by $R^+(j)$.

Our objective is to compute the probability $p(S, \mathcal{R})$ that every node of S is covered by an operating path of $\mathcal{R}$. The underlying tree structure allows this probability to be recursively calculated by traversing the tree from its leaves toward the root r. At the same time, the effects of routes confined to various subtrees of T need to be considered. To this end, we define the following quantities for all $j \in S$ and $i \in T_j$:

$g_i^j = \Pr(i$ not covered but all $x \in T'_i$ are covered using paths in $R^+(j))$,

$G_i^j = \Pr(\text{all } x \in T_i$ are covered using paths in $R^+(j))$.

The required coverage probability is then G_r^r. The RCP algorithm to be developed depends upon certain relations among the quantities defined above. A number of useful relations are given in the following theorem, where, by convention, any product taken over the empty set equals 1.

Theorem 7.7. *The following relations hold for $j \in S$ and $i \in T_j$:*

(a) $g_j^j = \prod \{q_k: R_k \in R(j)\} \cdot \prod \{G_x^x: x$ an immediate successor of $j\}$;
(b) *For* $i \neq j$, $g_i^j = g_i^x \cdot \prod \{q_k: R_k \in R(j), i \in R_k\}$, *where x is the root of the subtree in T'_j containing i;*
(c) $G_i^j = 1 - [g_i^j + (1 - \prod \{G_x^j: x$ an immediate successor of $i\})]$.

Proof.

(a) $g_j^j = \Pr(j$ not covered but all $y \in T'_j$ are covered using paths in $R^+(j))$
$= \Pr(\text{all paths in } R(j) \text{ fail}) \cdot \Pr(\text{each subtree rooted at an immediate successor } x \text{ of } j \text{ is covered using paths in } R^+(x))$
$= \prod \{q_k: R_k \in R(j)\} \cdot \prod \{G_x^x: x$ an immediate successor of $j\}$.

(b) For given $j \in S$ and $i \in T'_j$, let x be the root of the (unique) subtree in T'_j containing i.
$g_i^j = \Pr(i$ not covered but all $y \in T'_i$ are covered using paths in $R^+(j))$
$= \Pr(i$ not covered by a path in $R(j)$, i not covered by a path in $R^+(x)$, and all $y \in T'_i$ are covered using paths in $R^+(x))$
$= \prod \{q_k: R_k \in R(j), i \in R_k\} \cdot g_i^x$.

An algorithm for directed trees

(c) In the events below it is assumed that all paths are in $R^+(j)$.

$G_i^j = \Pr(\text{all } y \in T_i \text{ are covered})$
$= 1 - \Pr(\text{some } y \in T_i \text{ is not covered})$
$= 1 - \Pr(i \text{ not covered but all } y \in T_i' \text{ are covered, or } T_x \text{ is not covered for some immediate successor } x \text{ of } i)$
$= 1 - [\Pr(i \text{ not covered but all } y \in T_i' \text{ are covered}) + 1 - \Pr(\text{each } T_x \text{ is covered})]$
$= 1 - [g_i^j + (1 - \prod \{G_x^j : x \text{ an immediate successor of } i\})]$. □

Theorem 7.7 thus establishes the validity of the following algorithm for calculating the coverage probability using directed paths in a rooted directed tree.

RCP algorithm II. This algorithm calculates $p(S, \mathcal{R})$ for a rooted directed tree T with stops $S = \{1, 2, \ldots, m\}$ and directed paths $\mathcal{R} = \{R_1, R_2, \ldots, R_n\}$, where the routes R_k fail randomly and independently with probability q_k.

1. Place the stops $1, 2, \ldots, m$ in (reverse) topological order so that if y is a successor of x then $y < x$.

2. For $j = 1, 2, \ldots, m$
   (a) Compute g_j^j using Theorem 7.7(a).
   (b) For each $i \in T_j'$, compute g_i^j using Theorem 7.7(b).
   (c) For $i \in T_j$ (in reverse topological order) compute G_i^j using Theorem 7.7(c).

3. Output $p(S, \mathcal{R}) = G_m^m$.

Note that the nodes of a tree can be topologically ordered, as needed in Step 1, in $O(m)$ time. By performing an initial preprocessing step that requires $O(\sum |R_k|)$ work, Step 2(a) can be carried out using a total of $O(m)$ operations. This preprocessing step can also be used to accumulate the products of route failure probabilities needed in the formula of Theorem 7.7(b), so that each of the $O(m^2)$ products in Step 2(b) can then be carried out in constant time. Finally, the total work involved in Step 2(c) is $O(m^2)$. Thus the entire algorithm has a worst-case time complexity of $O(\max\{m^2, \sum |R_k|\})$. In the worst case the storage requirements are $O(m^2)$.

As an illustration of this algorithm, we consider again the subinterval problem of Section 7.1. This can be represented as an RCP on the rooted directed tree shown in Fig. 7.6. The four nodes (stops) have been

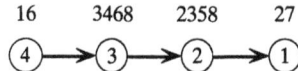

Fig. 7.6 Rooted directed tree for the example of Fig. 7.1.

ordered in reverse topological order, with the root being node 4. The routes involving each node are indicated next to the node. For simplicity, it is assumed that all routes R_k fail with probability $q_k = q$. Application of the above RCP algorithm produces

$j = 1$: $\quad g_1^1 = q; \quad G_1^1 = 1 - g_1^1 = 1 - q.$

$j = 2$: $\quad g_2^2 = q^2 G_1^1 = q^2 - q^3; \quad g_1^2 = g_1^1 q = q^2;$
$\quad\quad\quad G_1^2 = 1 - g_1^2 = 1 - q^2; \quad G_2^2 = 1 - [g_2^2 + 1 - G_1^2] = 1 - 2q^2 + q^3.$

$j = 3$: $\quad g_3^3 = q^3 G_2^2 = q^3 - 2q^5 + q^6; \quad g_1^3 = g_1^2 = q^2; \quad g_2^3 = g_2^2 q^2 = q^4 - q^5;$
$\quad\quad\quad G_1^3 = 1 - g_1^3 = 1 - q^2; \quad G_2^3 = 1 - [g_2^3 + 1 - G_1^3] = 1 - q^2 - q^4 + q^5;$
$\quad\quad\quad G_3^3 = 1 - [g_3^3 + 1 - G_2^3] = 1 - q^2 - q^3 - q^4 + 3q^5 - q^6;$

$j = 4$: $\quad g_4^4 = q^2 G_3^3 = q^2 - q^4 - q^5 - q^6 + 3q^7 - q^8;$
$\quad\quad\quad g_1^4 = g_1^3 = q^2; \quad g_2^4 = g_2^3 = q^4 - q^5; \quad g_3^4 = g_3^3 q = q^4 - 2q^6 + q^7;$
$\quad\quad\quad G_1^4 = 1 - g_1^4 = 1 - q^2; \quad G_2^4 = 1 - [g_2^4 + 1 - G_1^4] = 1 - q^2 - q^4 + q^5;$
$\quad\quad\quad G_3^4 = 1 - [g_3^4 + 1 - G_2^4] = 1 - q^2 - 2q^4 + q^5 + 2q^6 - q^7;$
$\quad\quad\quad G_4^4 = 1 - [g_4^4 + 1 - G_3^4] = 1 - 2q^2 - q^4 + 2q^5 + 3q^6 - 4q^7 + q^8.$

The expression for G_4^4 above gives the probability that all nodes of the path are covered by an operating subpath.

We conclude by remarking that the algorithm given here will be difficult to extend beyond the special case (*rooted* directed trees) considered in this section. As shown by Theorem 7.3, the RCP becomes #P-hard when we consider the analogous problem on arbitrary directed trees in which routes are directed paths.

7.5 Refinements and extensions

This chapter has examined the reliability covering problem, in which a given set of routes is used to cover a given set of stops. It has been shown that even for specialized cases (in which the size or structure of the routes is restricted), calculation of the probability of system coverage is difficult. Two solvable cases have, however, been identified, in which routes are defined with respect to an underlying tree, and polynomial algorithms have been developed for these cases. These two cases represent different generalizations of the situation in which routes are subpaths of a path, a problem previously analysed by Shanthikumar (1987) and by Hwang and Yao (1989). The important class of consecutive k-out-of-n systems, introduced in Chapter 6, turns out to be a special case of this special case (subpaths of a path).

In a *consecutive k-out-of-n failure system*, a set of ordered components $1, 2, \ldots, n$ is given in which each component i fails independently with probability q_i. The system functions unless some k (or more) consecutive components $j, j+1, \ldots, j+k-1$ all fail, where $k \le n$. Notice that a 1-out-of-n system corresponds to a system in which the n components are placed in series, whereas an n-out-of-n system corresponds to all components being placed in parallel. As an illustration of a situation in which a k-out-of-n system arises, suppose that n transmitters (at locations $l_1, l_2, \ldots, l_n$) are available to relay a message sent from a source location l_0 to a final location l_{n+1}. Because of the limited range of the transmission stations, each location l_j can forward a message only to the k nearby stations $l_{j+1}, l_{j+2}, \ldots, l_{j+k}$. If the n transmitters are subject to failure then a message can be successfully sent from l_0 to l_{n+1} unless some k (or more) successively numbered transmission stations are in a failed mode.

In general, a k-out-of-n failure system is coherent and thus can be completely described by its mincuts:

$$K_1 = \{1, 2, \ldots, k\}$$
$$\vdots$$
$$K_j = \{j, j+1, \ldots, j+k-1\}$$
$$\vdots$$
$$K_{n-k+1} = \{n-k+1, n-k+2, \ldots, n\}.$$

Again, this system can be viewed as an RCP having routes $R_1, R_2, \ldots, R_n$ and $n-k+1$ stops. Each route R_i is considered to fail with probability q_i. Route R_i contains those stops j such that $i \in K_j$: namely

$$R_1 = \{1\}$$
$$R_2 = \{1, 2\}$$
$$\vdots$$
$$R_{k+t} = \{t+1, t+2, \ldots, t+k\}, \quad 0 \le t \le n-2k+1$$
$$\vdots$$
$$R_{n-1} = \{n-k, n-k+1\}$$
$$R_n = \{n-k+1\}.$$

Notice that each of these routes is simply a subpath of the path $P: 1 \to 2 \to \cdots \to n-k+1$ connecting the stops in order. So this system is a special case of the linear subpath problem treated in Section 7.1 and can be efficiently solved using the methods of Sections 7.3 and 7.4.

The related *consecutive k-out-of-n success system* is considered to function whenever some k (or more) consecutive components all operate. Again, it is assumed that the components i operate independently with probability p_i. This is a coherent system $(C, \mathcal{P})$ with its $n-k+1$ minpaths given by $P_j = \{j, j+1, \ldots, j+k-1\}$ and it, too, can be expressed as a linear subpath covering problem. Namely, the n routes R_i correspond to the n components, and the $n-k+1$ stops correspond to the $n-k+1$ sets P_j. Route R_i contains precisely those stops j for which $i \in P_j$. There is a one-to-one correspondence between route covers of this RCP and failure sets of the coherent system. To calculate the probability that $(C, \mathcal{P})$ does *not* operate, we consider each route R_i to *fail* with probability p_i and then determine $p(S, \mathcal{R})$. This latter probability is easily calculated using either of the two RCP algorithms presented earlier.

The various problems treated in this chapter involve covering the nodes of a given network G by certain routes, viewed as subsets of nodes. There are also situations in which it is of interest to cover both the nodes and edges of G by routes. For example, if the routes represent street segments patrolled by a city's police cars, then it is important to carry out surveillance along the entire length of the streets (edges) as well as at the intersections (nodes). In general, this type of problem can be modelled by inserting a new degree-two node along each edge of G, forming a new network G'. Any specified route for G can then be modified to include any such intermediate nodes corresponding to the edges it also serves. As a result, the probability that the nodes and edges of G are covered by the original routes is the same as the probability that all nodes of G' are covered by the modified (node subset) routes. The two polynomial algorithms presented here can then be applied to solve certain cases of this mixed node/edge covering problem.

7.6 Chapter notes

A considerable literature exists on consecutive k-out-of-n systems. Chiang and Chiang (1986) and Chiang and Niu (1981) discuss various applications of this type of coherent system. Efficient recursive algorithms for computing the reliability of consecutive k-out-of-n systems are given by Hwang (1982) and Shanthikumar (1982). Kossow and Preuss (1989) derive a topological formula for terms of the inclusion–exclusion expansion applied to such systems. Again, the coefficients of each noncancelling term in the reduced expression turn out to have a predictable ± 1 value. A more general result of this type appears in Shier (1990).

Shanthikumar (1987) and Hwang and Yao (1989) provide natural generalizations of consecutive k-out-of-n systems to certain types of 'consecutively connected' systems. Ball et al. (1989) introduce the reliability covering problem and develop algorithms for the two generalizations of consecutively connected systems treated in this chapter.

8
State-space approximation

The emphasis in the previous chapters has been on assessing the reliability of a binary system, whose components (nodes and/or edges) can be in either of two states (operating or failed). However, reliability is only one of a number of stochastic performance measures that might be of practical interest. For example, the number of node pairs that can communicate at a random instant might be more informative than simply whether or not all node pairs can communicate. Alternatively, we might be interested in the 'fault tolerance' of the system, measured by the expected number of disjoint paths joining two specified nodes. Since the exact calculation of these and other realistic performance measures is known to be #P-hard, there is reason to investigate methods for approximating such measures. While it might be thought that obtaining a good approximation to the performance of a network might be substantially easier than computing its exact value, this does not appear to be the case. For instance, Ball (1980) has shown that obtaining an approximation to network reliability that is guaranteed to be within 100ε percent of the exact value is still #P-hard for a variety of reliability measures. The status of obtaining an approximation guaranteed to be within a prescribed absolute error ε of the exact value is at present unresolved.

The purpose of this chapter is to re-examine the state-space enumeration method, described in Chapter 2, and examine how it can be modified to approximate general performance measures defined for a binary system. The overall strategy involves generating a relatively small number of system states that encompass a relatively large proportion of the total probability. Specifically, we discuss the problem of generating the states of a system in order of nonincreasing probability. Such an ordering ensures that maximum coverage of the state space (in terms of probability) will be obtained for a specified number of generated states. It will be seen that an elegant algebraic structure, a distributive lattice, underlies this generation problem. Moreover, this algebraic formulation naturally leads to an algorithm for generating, in order, the states of the given system. In typical cases the most probable states are produced by examining only a tiny portion of the entire state space. Once the generation procedure has been carried out, various bounds on performance measures for the system are then easily obtained. An attractive feature of this approach is that lower and upper bounds on system performance can be produced at each step, so the process can be continued until the bounds become sufficiently close. The chapter closes with a discussion of how the generation of most probable states can be carried out in certain 'multistate' systems.

8.1 Lattice structure

Consider a binary system involving the set of components $E = \{1, 2, \ldots, n\}$. Each component $i \in E$ fails independently with probability $q_i = 1 - p_i$. Without loss of generality, it can be assumed that each component reliability $p_i \geq \frac{1}{2}$, since otherwise the analysis can proceed using $1 - p_i$ in place of p_i. Also, the components are assumed to be numbered from least reliable to most reliable, giving

$$\tfrac{1}{2} \leq p_1 \leq p_2 \leq \cdots \leq p_n \leq 1. \tag{8.1}$$

Consequently, the ratios $R_i = q_i/p_i$ are placed in nonincreasing order:

$$1 \geq R_1 \geq R_2 \geq \cdots \geq R_n \geq 0. \tag{8.2}$$

Each state of the system is defined by the subset $X \subseteq E$ of *failed* components occurring in that system state. The set of all system states is denoted by $\mathscr{S} = \mathscr{S}_n$. Assuming independent failures, the probability $p(X)$ of state $X \in \mathscr{S}$ is then given by

$$p(X) = \prod_{i \in \bar{X}} p_i \prod_{j \in X} q_j = \prod_{i=1}^{n} p_i \prod_{j \in X} R_j = \left(\prod_{i=1}^{n} p_i\right) R(X), \tag{8.3}$$

where $R(X)$ is the *R-value* of state X. The special case when $X = \emptyset$ (all n components being operational) is assigned the R-value 1, consistent with eqn (8.3).

The approach taken here is to generate the states X in order of nonincreasing probability $p(X)$. That is, the states of the system are required in the order $X_1, X_2, \ldots$, where $p(X_i) \geq p(X_j)$, for $i < j$. In a typical scenario we would like to generate only $k \ll 2^n = |\mathscr{S}|$ states, so the choice of the k most probable states yields the maximum sum of state probabilities, or *coverage probability*, for any k selected states. It would be fortunate indeed if the coverage probability associated with the k generated states were close to 1. However, if all $p_i = \frac{1}{2}$, then every state has equal probability and, in order to obtain coverage probability π, that same proportion of the system states needs to be generated. In this case, there is little point in employing the state-space approximation approach.

Fortunately, in a number of instances it turns out that a relatively small proportion of the system states accounts for a large proportion of the total probability. To illustrate this phenomenon, suppose, for simplicity, that all components have the same reliability $p_i = p$. Table 8.1 displays

Lattice structure

Table 8.1 States needed to achieve $\pi = 0.90$ coverage

	% of states		
	$p = 0.85$	$p = 0.9$	$p = 0.95$
$n = 10$	12.7	4.88	1.07
$n = 15$	4.54	1.28	0.22
$n = 20$	1.60	0.30	0.018
$n = 25$	0.56	0.045	0.003
$n = 30$	0.19	0.013	0.0003

the percent of states that needs to be generated (in most probable order) to achieve coverage probability $\pi = 0.90$. For example, in a 25-component system with $p = 0.9$, only 0.045% of the system states needs to be generated to achieve the specified coverage. As seen in the table, this proportion dramatically decreases as n and p increase. Thus there is hope for being able to approximate the performance of binary systems if the individual components are sufficiently reliable. Fortunately, this is frequently the case in practice and in such situations the generation of most probable states can provide useful approximations.

Suppose, then, that the k most probable states $X_1, X_2, \ldots, X_k$ have been successfully generated. With each system state $X \in \mathcal{S}$ there is also defined some real-valued performance measure $\mu(X)$. For example, $\mu(X)$ might indicate whether all nodes are connected, the number of node pairs able to communicate, or the number of disjoint s-t paths in system state X. Of interest in this context is the *expected performance* of the system $\bar{\mu} = \sum p(X)\mu(X)$, averaged over all $X \in \mathcal{S}$. Using the k generated states then gives the following lower and upper bounds on $\bar{\mu}$:

$$\bar{\mu}_L = \sum_{i=1}^{k} p(X_i)\mu(X_i) + \left[1 - \sum_{i=1}^{k} p(X_i)\right]\mu_L, \quad (8.4)$$

$$\bar{\mu}_U = \sum_{i=1}^{k} p(X_i)\mu(X_i) + \left[1 - \sum_{i=1}^{k} p(X_i)\right]\mu_U, \quad (8.5)$$

where μ_L and μ_U are, respectively, known lower and upper bounds on the performance in any system state. For the case of binary systems, we can certainly take $\mu_L = 0$ and $\mu_U = 1$.

Our subsequent attention will thus be directed to the problem of generating states in order of nonincreasing probability. It will be seen that simply knowing the ordinal relation conveyed by (8.1) provides considerable information about the relative probabilities of the various

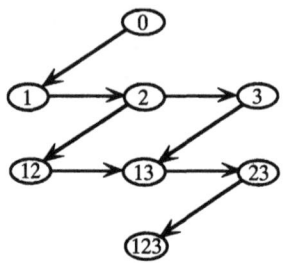

Fig. 8.1 Hasse diagram for the three-component system.

states $X_i \in \mathcal{S}$. To clarify this connection, define for any two states X_i, X_j, the order relation $X_i \geqslant X_j$, if $p(X_i) \geq p(X_j)$ holds for *all* values p_r satisfying (8.1). It is then straightforward to show the following.

Property 8.1. *The set $\mathcal{S}$ of all states forms a partially ordered set $(\mathcal{S}, \geqslant)$.*

We can represent the partially ordered set $(\mathcal{S}, \geqslant)$ by a *graph* whose elements are joined by directed *arcs*. Each state $X \in \mathcal{S}$, containing the failed components $i_1, i_2, \ldots, i_r$, is labeled by $i_1 i_2 \cdots i_r$ with $i_1 < i_2 < \cdots < i_r$. The maximal element $X = \emptyset$ (all components working) is designated by the label 0. Note that, by eqn (8.3), comparing $p(X_i)$ with $p(X_j)$ for all p_r satisfying (8.1) is equivalent to comparing $R(X_i)$ with $R(X_j)$ for all R_r satisfying (8.2), so that the ordering of R-values in (8.2) completely determines this directed graph. Consider, for example, the partially ordered set $(\mathcal{S}_3, \geqslant)$. We see that $2 \geqslant 3 \geqslant 13$ holds since $R_2 \geq R_3 \geq R_1 R_3$. The Hasse diagram for $(\mathcal{S}_3, \geqslant)$ is shown in Fig. 8.1, while the Hasse diagram for the case of four components is shown in Fig. 8.2.

In general, the Hasse diagram for n components consists of 2^n elements, one for each state of the system. The 2^n elements are arranged

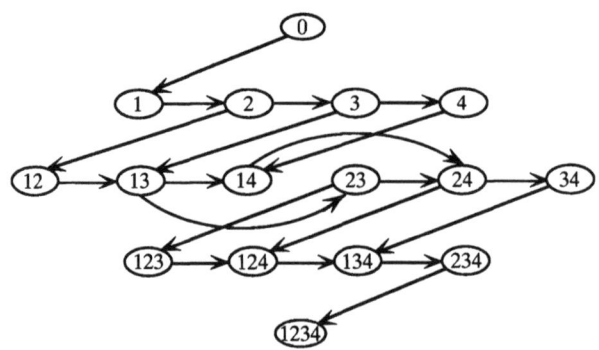

Fig. 8.2 Hasse diagram for the four-component system.

into $n+1$ levels such that level $j = 0, \ldots, n$ contains $\binom{n}{j}$ elements. Each element on level j corresponds to a state with exactly j failed components. The arcs of the Hasse diagram can be separated into two different categories: arcs within a given level and arcs between two consecutive levels. Within each level j ($j = 1, \ldots, n-1$), an arc extends from element $k_1 \cdots k_{i-1} k_i \cdots k_j$ to element $k_1 \cdots k_{i-1} k_i + 1 \cdots k_j$ provided $k_i + 1$ and k_{i+1} are distinct and $k_i + 1 \le n$. Between two consecutive levels j and $j+1$ ($j = 0, \ldots, n-1$), an arc extends from each element $k_1 k_2 \cdots k_j$ on level j with $k_1 \ne 1$ to the element $1 k_1 k_2 \cdots k_j$ on level $j+1$.

An interesting feature of the Hasse diagram based on $n \ge 2$ components is that it contains two copies of the Hasse diagram on $n-1$ components. One of the copies consists of all elements in which component n is operating: this copy is precisely the Hasse diagram for the $(n-1)$-component system. The second copy contains all elements in which component n has failed. The labels in this copy simply have the integer n adjoined to the end of each label in the first copy. It is straightforward to verify that there are 2^{n-2} arcs joining these two copies. The Hasse diagram of the four-component system shown in Fig. 8.3 illustrates the above duplication feature. The two copies of the Hasse diagram on three components are shown with heavy lines, while the $2^{4-2} = 4$ arcs joining the copies are shown as dashed lines.

If f_n denotes the number of arcs in the Hasse diagram for n components, then this duplication property shows that f_n satisfies $f_n = 2f_{n-1} + 2^{n-2}$, $n \ge 2$. Solving this recursion with $f_1 = 1$ then yields the following.

Property 8.2. *The number of arcs in the Hasse diagram for n components is $(n+1)2^{n-2}$.*

This same partially ordered set $(\mathscr{S}_n, \ge)$ arises in other contexts (Proctor 1982a, b; Stanley 1980) and certain established properties of this

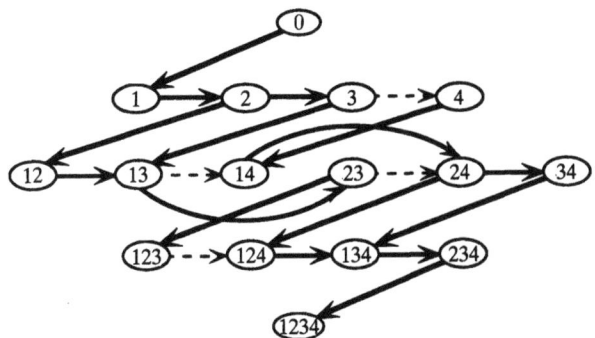

Fig. 8.3 Duplication feature in the four-component system.

structure are now briefly mentioned. First, the partially ordered set forms a special type of lattice, a distributive lattice. Essentially it is a sublattice of the Cartesian product of n chains and hence is distributive (Stanley 1980). Moreover, the lattice can be *ranked* or *graded*: namely, its elements can be decomposed into subsets $P_0, P_1, \ldots, P_h$ such that arcs of the Hasse diagram join elements only in consecutive sets P_r. The rank of any element is simply the sum of the integers comprising its label. Thus element 0 has rank 0, element $12 \cdots n$ has rank $n(n+1)/2$, and so the lattice has a *height* (maximum rank) $h = n(n+1)/2$. In an n-component system with height h, let $\mathbf{r} = (r_0, r_1, \ldots, r_h)$ be its rank vector, with $r_i = |P_i|$ signifying the number of elements having rank i. As shown by Proctor (1982a) and Stanley (1980), the rank vector is symmetric and unimodal. For example, the Hasse diagram for a three-component system has $h = 6$ and the rank vector $\mathbf{r} = (1, 1, 1, 2, 1, 1, 1)$, which is clearly symmetric and unimodal. The rank vector for the system $\mathscr{S}_4$ is $\mathbf{r} = (1, 1, 1, 2, 2, 2, 2, 2, 1, 1, 1)$.

Once the state space has been identified as a partially ordered set, it is not difficult to formulate an algorithm for generating, in order, the most probable states of the system. This algorithm, described in the next section, conceptually works on the Hasse diagram. Because the Hasse diagram contains 2^n elements, its direct construction is undesirable. In fact, the algorithm of the next section considers only a small portion of the Hasse diagram which is then extended in a local fashion from iteration to iteration, so that it is possible to carry out the needed steps without examining the entire state space $\mathscr{S}_n$.

8.2 An algorithm for state generation

This section describes an algorithm, based on the Hasse diagram, for generating the most probable states in order. Certain properties of the algorithm are also investigated. In addition, we examine a worst-case upper bound on the complexity of this algorithm, which turns out to be related to a certain invariant of the underlying partial order.

In the Hasse diagram for a partially ordered set, let the *indegree* of an element be the number of elements (or *predecessors*) that cover the given element: i.e., the number of arcs entering that element in the Hasse diagram. Any (finite) partially ordered set contains at least one element with indegree 0. In our particular case, $(\mathscr{S}, \geq)$ contains a single such element (the one labelled 0) and it corresponds to the most probable state of the system. When this element is removed from the Hasse diagram, there will be at least one element in the reduced Hasse diagram with indegree 0. More generally, when the k most probable elements have been removed from the Hasse diagram, there will remain some set

$\mathscr{C}$ of elements having indegree 0 in the reduced Hasse diagram. Any two such elements $X, Y \in \mathscr{C}$ must be incomparable in the original partial order and thus are candidates for selection as the next most probable state. The partial information embodied in (8.1) is not sufficient to completely order, in advance, the probabilities of states represented in $\mathscr{C}$. However, by comparing R-values of the candidates, the next most probable state (selected from $\mathscr{C}$) can be determined.

An algorithm for generating the most probable states in order follows directly. At each step of the algorithm we remove from $\mathscr{C}$ an element X with largest R-value and then update the candidate set $\mathscr{C}$, since the removal of X (including its arcs) may create new elements with indegree 0. In particular, any *successors* Y of X (having X as a predecessor) will have their indegree reduced by 1. In order to avoid scanning the entire partial order (with 2^n states), it is useful to maintain an 'active set' $\mathscr{A}$, which contains those elements in the Hasse diagram having a predecessor that has already been removed from the Hasse diagram. An element is transferred from the set $\mathscr{A}$ to the candidate set $\mathscr{C}$ whenever its indegree becomes 0. This process of successively removing elements from $\mathscr{C}$ and updating the relevant sets is continued until some stopping criterion is satisfied. For our purposes, termination is governed by achieving a specified coverage probability π. The algorithm GENERATE implements this approach to identify the most probable states in order of nonincreasing probability.

Algorithm GENERATE. Given certain component probabilities $p_1, \ldots, p_n$ ordered as in (8.1), this algorithm generates states in order of nonincreasing probability until the specified coverage π is obtained.

1. [Initialization]
   $\mathscr{C} := \{0\}$; $\mathscr{A} := \varnothing$; sum $:= 0$;
2. [Iterative Step]
   while (sum $< \pi$) do
   delete the element $X \in \mathscr{C}$ having the largest R-value;
   sum $:=$ sum $+ p(X)$;
   output X;
   find successors of X, place them on $\mathscr{A}$ (if not already
       present), and update their indegrees;
   remove all elements with indegree 0 from $\mathscr{A}$ and transfer
       them to $\mathscr{C}$.

The operation of this algorithm is illustrated using a six-component example with $p_1 = 0.75$, $p_2 = 0.8$, $p_3 = 0.8$, $p_4 = 0.85$, $p_5 = 0.9$, $p_6 = 0.95$. Table 8.2 shows, for this example (run until a coverage of $\pi = 0.90$ is achieved), the sizes of the candidates and active sets, along with the most

Table 8.2 Sizes of sets used by GENERATE in a six-component example

Iteration	$|\mathscr{C}|$	$|\mathscr{A}|$	Most probable state
—	—	—	0
1	1	0	1
2	1	0	2
3	2	0	3
4	2	1	4
5	2	2	5
6	2	3	12
7	2	2	13
8	3	1	23
9	3	2	14
10	4	0	6
11	3	1	24
12	3	3	34
13	2	5	15

probable state output at each iteration. In this small example, 14 states (out of the 64 states comprising the state space) were needed to achieve the desired coverage probability. Moreover, the various sets required by GENERATE are seen to stay quite modest in size: the size of the candidate set never exceeded 4 and the size of the active set never exceeded 5. This example illustrates the general finding that, in practice, only a small portion of the state space requires explicit examination. Theoretical reasons for this behavior will be discussed later.

We now establish certain properties of the sets used in this algorithm. Let $\mathscr{D}$ denote the set of elements deleted from the partially ordered set at some step of the algorithm. Thus the states in $\mathscr{D}$ have already been correctly generated in order. The 'neighbourhood' of $\mathscr{D}$ in $(\mathscr{S}, \geqslant)$ is defined by $\Gamma(\mathscr{D}) = \{X \notin \mathscr{D}: X \text{ has a predecessor } Y \in \mathscr{D}\}$ so $\Gamma(\mathscr{D}) = \mathscr{C} \cup \mathscr{A}$. Recall that a set $\mathscr{K} \subseteq \mathscr{S}$ is termed *convex* if, for any $X, Z \in \mathscr{K}$ and $Y \in \mathscr{S}$ with $X \geqslant Y \geqslant Z$, we have $Y \in \mathscr{K}$. The following two results are now demonstrated.

Lemma 8.3. *The set $\mathscr{D}$ is a convex subset of $\mathscr{S}$.*

Proof. Suppose $X, Z \in \mathscr{D}$ with $X \geqslant Y \geqslant Z$. Since Y is more probable than Z it must be removed before Z. Because $Z \in \mathscr{D}$ it follows that $Y \in \mathscr{D}$. □

Theorem 8.4. *At any step of the algorithm $|\Gamma(\mathscr{D})| \leqslant |\mathscr{D}|$.*

Proof. We establish a one-to-one mapping from $\Gamma(\mathcal{D})$ to $\mathcal{D}$. Suppose that element X is in $\Gamma(\mathcal{D})$. Then $X = i_1 i_2 i_3 \cdots i_m \cdots i_t$ has a predecessor $Y \in \mathcal{D}$, say $Y = i_1 i_2 i_3 \cdots i_m - 1 \cdots i_t$. If there are several predecessors, the choice can be made arbitrarily. Also, if $m = 1$ and $i_m = 1$, we interpret this to mean $Y = i_{m+1} \cdots i_t$. Now define the mapping $\psi: \Gamma(\mathcal{D}) \to \mathcal{D}$ using $\psi(X) = i_1 i_2 i_3 \cdots i_{m-1} i_{m+1} \cdots i_t$. Notice that $0 \geqslant \psi(X) \geqslant Y$, where element 0 is the most probable state. Since $0, Y \in \mathcal{D}$ then the convexity of $\mathcal{D}$ (Lemma 8.3) yields $\psi(X) \in \mathcal{D}$.

Now suppose that there is some $X' \in \Gamma(\mathcal{D})$, $X' \neq X$, with $\psi(X') = \psi(X)$. Then X' must have a predecessor $Y' \in \mathcal{D}$, which we assume has the form

$$Y' = i_1 i_2 i_3 \cdots i_{m-1} i_{m+1} \cdots i_{r-1} j i_r \cdots i_t.$$

(A similar argument governs other placements of the index j.) Since the indices satisfy $i_m < i_{m+1} < \cdots < i_{r-1} < j < i_r < \cdots < i_t$, it can be verified that $Y \geqslant X \geqslant Y'$ holds. Again, by the convexity of $\mathcal{D}$ and the fact that $Y, Y' \in \mathcal{D}$, it follows that $X \in \mathcal{D}$, a contradiction. Thus the mapping is one-to-one and the stated result follows. □

If $k = |\mathcal{D}|$ states are needed to achieve the desired coverage probability, then the above theorem assures us of a modest growth in the total number of states that must be consulted (and stored) during the search. Specifically, Theorem 8.4 shows that $|\mathscr{C} \cup \mathscr{A}| \leq k$ and thus at most $2k$ states need to be examined and processed by the algorithm. Moreover, as seen earlier, if the component probabilities are all sufficiently large (a situation regularly encountered in practice), then k itself will be small relative to the 2^n system states. In such situations, state-space approximation could be a viable approach whereas complete enumeration would be out of the question.

The computational complexity of algorithm GENERATE, both theoretically and empirically, is determined in large part by the effort needed to process the candidate set $\mathscr{C}$. While Theorem 8.4 provides the upper bound $k = |\mathcal{D}|$ on the maximum size of the set $\mathscr{C}$, there is an alternative upper bound related to the structure of the partially ordered set $(\mathscr{S}_n, \geqslant)$. Notice that all elements in $\mathscr{C}$ must be incomparable in the partial order since each has indegree 0 in the reduced Hasse diagram. In other words, the elements of $\mathscr{C}$ form an *antichain* in the partially ordered set: a set of mutually incomparable elements in the partial order. Thus an upper bound on the maximum size of $\mathscr{C}$ is the size α of the largest antichain in $(\mathscr{S}_n, \geqslant)$.

As indicated earlier, the elements of this partially ordered set can be partitioned into sets P_i of rank i. Moreover, the partially ordered set is known to be *Sperner* (Stanley 1980), meaning that the size of the largest antichain is the same as the maximum size of the sets P_i. Since the partially ordered set is rank unimodal and rank symmetric, a maximum-sized antichain occurs for a set P_i at the half-height δ of the lattice: namely, at rank $\delta = \left\lfloor \dfrac{h}{2} \right\rfloor = \left\lfloor \dfrac{n(n+1)}{4} \right\rfloor$. As a result, we obtain the upper bound $\alpha = |P_\delta|$ on the maximum possible size of the candidate set $\mathscr{C}$. Since the rank of the element $i_1 i_2 \cdots i_t$ is simply the sum of its constituent integers i_j, another way of expressing this upper bound α is as the number of partitions of the integer δ into distinct parts, none of which exceeds n. Equivalently, α can be determined as the coefficient of x^δ in the generating function $(1+x)(1+x^2)(1+x^3) \cdots (1+x^n)$.

8.3 Critical sets

The previous section developed an algorithm for generating the states $X_1, X_2, \ldots$ of a system in order of nonincreasing probability. For the case of highly reliable components, relatively few such generated states will typically account for a disproportionately large fraction of the total probability. In order to produce upper and lower bounds on the expected performance $\bar{\mu}$ of the system via (8.4) and (8.5), $\mu(X_i)$ must be evaluated for each of the k generated states X_i. Since direct evaluation of each $\mu(X_i)$ can be quite time-consuming, it is desirable to have a method that predicts the performance value of new states based on the performance value of existing states. This section treats the specific case of coherent binary systems and shows that, by suitable preprocessing, the performance of each state can be easily deduced from that of a predecessor state.

Recall that a coherent binary system is defined with respect to a set $E = \{1, \ldots, n\}$ of binary components, each of which operates independently with some known probability. A structure function $\Phi(S)$ is defined for each set $S \subseteq E$ of working components. Namely, $\Phi(S) = 1$ if the system operates and $\Phi(S) = 0$ if the system fails, when the components in S operate and those in $E - S$ fail. The structure function of a coherent system is a monotone nondecreasing set function; see eqn (6.3). The collection of minpaths for the coherent system will be denoted $\mathscr{P}$, while the collection of mincuts will be denoted $\mathscr{K}$. In addition, we denote by $\bar{\mathscr{P}}$ the set of complements of minpaths: $\bar{\mathscr{P}} = \{E - P : P \in \mathscr{P}\}$.

Because the state generation method described in this chapter works with states X representing *failed* components, it will be notationally convenient to introduce the complementary structure function Ψ defined by $\Psi(X) = \Phi(E - X)$. Hence $\Psi(X)$ is a monotone nonincreasing function on sets of failed components:

$$A \subseteq B \Rightarrow \Psi(A) \geq \Psi(B). \tag{8.6}$$

Notice that the generation algorithm of Section 8.2 produces, at each step, certain successor states Y from a selected state X (each state is described by its failed components). Either Y is obtained from X by exchanging the failed component $i+1$ for the failed component i, giving $Y = X - \{i\} + \{i+1\}$, or it is obtained from X by adding the failed component 1, giving $Y = X + \{1\}$. We would like to predict whether $\Psi(Y) = \Psi(X)$ or $\Psi(Y) = 1 - \Psi(X)$ holds when exchanging component j for component i, or adding a new component i. For notational simplicity, the set $W \cup \{i\}$, with $i \notin W$, will be denoted $W + i$. A *critical (i, j) operating set* is then defined as a set $W + i$ such that $\Psi(W + i) = 1$ and $\Psi(W + j) = 0$. In other words, replacing failed component i by failed component j changes an operating system to a failed system.

Suppose that $W + i$ is a critical (i, j) operating set and $Z \supseteq W$ satisfies $\Psi(Z + i) = 1$. Then $Z + i$ will also be a critical (i, j) operating set. This assertion follows since $Z \supseteq W$ implies $Z + j \supseteq W + j$ and so, by (8.6), $\Psi(Z + j) \leq \Psi(W + j) = 0$, whence $\Psi(Z + j) = 0$. As a consequence, it will be sufficient to study the *minimal* critical (i, j) operating sets. Knowledge of these minimal critical sets then implicitly describes all critical sets. The following theorem provides a characterization of the minimal critical (i, j) operating sets.

Theorem 8.5. *$B + i$ is a minimal critical (i, j) operating set if and only if there exists $A \supseteq B$ such that $A + i \in \bar{\mathcal{P}}$ and $B + j \in \mathcal{K}$.*

Proof. ($\Rightarrow$) Suppose $B + i$ is a minimal critical set. Since $\Psi(B + i) = 1$ there exists some $A \supseteq B$ such that $A + i \in \bar{\mathcal{P}}$. Also, $\Psi(B + j) = 0$ and so there exists some mincut $C \in \mathcal{K}$ with $B + j \supseteq C$. It is seen that $j \in C$ since otherwise $B \supseteq C$, $\Psi(B + i) \leq \Psi(B) \leq \Psi(C) = 0$, contradicting $\Psi(B + i) = 1$. Thus we can write $C = Z + j \in \mathcal{K}$ with $\Psi(Z + j) = 0$ and $Z \subseteq B$. Then $Z + i \subseteq B + i$, $\Psi(Z + i) \geq \Psi(B + i) = 1$, and so $\Psi(Z + i) = 1$. Thus $Z + i$ is a critical (i, j) operating set with $Z + i \subseteq B + i$; by the minimality of $B + i$ it follows that $Z = B$ and so $B + j \in \mathcal{K}$.

($\Leftarrow$) Since $B \subseteq A$, $B + i \subseteq A + i$, $\Psi(B + i) \geq \Psi(A + i) = 1$, and so $\Psi(B + i) = 1$. Moreover, $B + j$ is a mincut so $\Psi(B + j) = 0$, meaning that $B + i$ is a critical (i, j) operating set. We claim that $B + i$ is also minimal. If not, then let $Z + i$ be any critical (i, j) operating set with $Z \subset B$. Now $Z + j \subset B + j$ and, by the minimality of the mincut $B + j$, it follows that $\Psi(Z + j) = 1$, contradicting the criticality of $Z + i$. $\square$

We can also study the effect of adding a new component i to a given set of components. A *critical operating set for adding i* is thus defined as a set W such that $\Psi(W) = 1$ and $\Psi(W + i) = 0$. The next theorem provides a characterization of the minimal such sets, relative to adding component i.

Theorem 8.6. *W is a minimal critical operating set for adding i if and only if $W + i \in \mathcal{K}$.*

Proof. ($\Rightarrow$) Suppose W is a minimal critical set for adding i. Since $\Psi(W + i) = 0$, $W + i$ contains some mincut $C \in \mathcal{K}$. It follows that $i \in C$ since otherwise $W \supseteq C$, $\Psi(W) \leq \Psi(C) = 0$, contradicting $\Psi(W) = 1$. Thus $C = Z + i \in \mathcal{K}$ with $\Psi(Z + i) = 0$ and $Z \subseteq W$. Also, $\Psi(Z) = 1$ follows from the minimality of the mincut $Z + i$, whence Z is a critical set for component i. Since W is assumed to be a minimal critical set, we must have $W = Z$ and thus $W + i \in \mathcal{K}$.

($\Leftarrow$) Since $W + i \in \mathcal{K}$, $\Psi(W + i) = 0$ and $\Psi(W) = 1$. W is then a critical operating set. To show that it is minimal, suppose Z is some critical operating set for i with $Z \subset W$. Now $Z + i \subset W + i \in \mathcal{K}$ and so $\Psi(Z + i) = 1$, contradicting the presumed criticality of Z. □

For sake of completeness, we record the corresponding result for critical (i, j) *failure sets*: that is, sets $W + i$ such that $\Psi(W + i) = 0$ and $\Psi(W + j) = 1$.

Theorem 8.7. *$A + i$ is a maximal critical (i, j) failure set if and only if there exists $B \subseteq A$ such that $B + i \in \mathcal{K}$ and $A + j \in \bar{\mathcal{P}}$.*

Suppose that sets A, B have been determined with $A \supseteq B$, $A + i \in \bar{\mathcal{P}}$, and $B + j \in \mathcal{K}$. Theorem 8.5 shows that $B + i$ is a minimal critical (i, j) operating set and Theorem 8.7 shows that $A + j$ is a maximal critical (j, i) failure set. A final result concerns removal of component i from a given set W. A *critical failure set for removing* i is any set W containing i for which $\Psi(W) = 0$ and $\Psi(W - i) = 1$.

Theorem 8.8. *W is a maximal critical failure set for removing i if and only if $W - i \in \bar{\mathcal{P}}$.*

To illustrate these concepts and results, consider the two-terminal network in Fig. 8.4. The minpaths of this system are the s-t paths $\mathcal{P} = \{16, 1347, 138, 2456, 247, 28\}$ so that $\bar{\mathcal{P}} = \{234578, 2568, 24567, 1378, 13568, 134567\}$. The mincuts of this system are the s-t cutsets

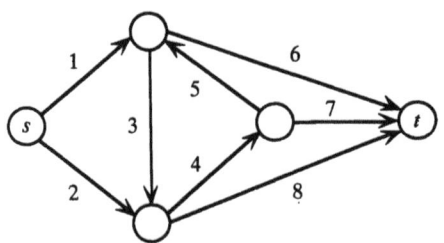

Fig. 8.4 Critical sets in an eight-component system.

$\mathcal{K} = \{12, 148, 1578, 236, 468, 678\}$. Because $2568 \in \bar{\mathcal{P}}$ and $468 \in \mathcal{K}$, the above results show that 268 is a minimal critical (2, 4) operating set and 4568 is a maximal critical (4, 2) failure set. Also, since $148, 468 \in \mathcal{K}$, then 18, 68 are the minimal critical operating sets for adding component 4. The maximal critical failure sets for removing component 4 are 24568, 13478, 134568.

Knowledge of these critical sets can be used in conjunction with algorithm GENERATE of Section 8.2 to approximate the expected performance $\bar{\mu}$ of a coherent system. Initially we are assured that $\Psi(\emptyset) = 1$ by the assumption of coherence. Suppose that $\Psi(X)$ is known when state X is removed from the candidate set. Then, by applying Theorems 8.5–8.7, the performance value $\Psi(Y)$ will be determined for each of the successors Y of X. For example, if $\Psi(X) = 1$ and $Y = X + \{1\}$ then $\Psi(Y) = 1$ unless Y contains some minimal critical operating set for adding component 1. If $\Psi(X) = 1$ and $Y = X - \{i\} + \{i + 1\}$, then $\Psi(Y) = 1$ unless Y contains some minimal critical $(i, i + 1)$ operating set. On the other hand, if $\Psi(X) = 0$ and $Y = X - \{i\} + \{i + 1\}$, then $\Psi(Y) = 0$ unless Y is contained in some maximal critical $(i, i + 1)$ failure set. Of course, if $\Psi(X) = 0$ and $Y = X + \{1\}$ then $\Psi(Y) = 0$. It then follows that every state X will have its performance value $\Psi(X)$ available as it is generated, without any need for explicit computation of $\Psi(X)$. This approach is particularly attractive when evaluating the performance of a state is computationally expensive or when there are a large number of states k that require evaluation.

8.4 Multistate systems

The previous sections of this chapter have studied state-space approximation applied to binary systems: systems whose components can assume one of two states (operating or failed). However, in many practical settings the system components can realize more than two modes of operation. For example, the components of a communication network can undergo gradual deterioration rather than simply fail. Likewise, the operation of a power distribution system can reflect varying levels of generating and transmission capacity. In these cases the system is more realistically modelled by allowing components to assume a (finite) number of states, each corresponding to a different level of operating efficiency. In such a multistate system, we are again interested in the stochastic behavior of various performance measures. The #P-hardness of computations in the special case of binary systems shows that exact calculation of these performance measures remains difficult when components can occupy more than two states. Thus there is reason to investigate the state-space approximation approach for multistate systems.

A *multistate system* consists of n components, with component i operating in any one of m_i distinct *modes* $0, 1, \ldots, m_i - 1$. All components are assumed to behave independently, with the probability that component i is in mode j being denoted by $p_{ij} > 0$. Hence

$$\sum_{j=0}^{m_i-1} p_{ij} = 1 \quad \text{for } i = 1, 2, \ldots, n.$$

The modes for each component i are assumed to be ordered so that

$$p_{ij} \geq p_{ik} \quad \text{for } j < k. \tag{8.7}$$

A *state* of the system can be represented by an n-vector X whose ith element $X[i]$ indicates the mode of component i. The probability of state X is given by

$$p(X) = \prod_{i=1}^{n} p_{i,X[i]}.$$

The entire state space, denoted by $\mathscr{S}$, has cardinality $\prod_{i=1}^{n} m_i$. Because there are an exponential number of states, our approach will be to enumerate states $X_0, X_1, \ldots$ in order of nonincreasing probability. In view of (8.7), state $X_0 = (0, 0, \ldots, 0)$ is the most probable state. Typically, just enough states need to be enumerated so that the sum of probabilities of the generated states is at least π, some specified coverage probability.

Empirical evidence (Chiou and Li 1986; Shier *et al.* 1990) suggests that in many practical multistate problems most of the state probability is captured in relatively few states. Thus the generation of most probable states might be a reasonable method for approximating the performance of the system. Unlike the case for binary components, there is no longer a simply described lattice structure present to guide the efficient generation of states. Several algorithms have, however, been proposed for generating the states of a multistate system in order of nonincreasing probability.

The algorithm developed by Chiou and Li (1986) begins by sorting in decreasing order all the p_{ij} terms with $j \neq 0$. Also, the initial set of states $\mathscr{S}_0 = \{X_0\}$ is defined. At each step k of the algorithm, the next highest probability p_{ij} in order is processed and used to expand the existing set of examined states $\mathscr{S}_k$. Specifically, each state $X \in \mathscr{S}_k$ having $X[i] = 0$ is used to generate a new state with $X[i] = j$. These new states are merged with those currently in $\mathscr{S}_k$ to obtain $\mathscr{S}_{k+1}$, and the process is continued until enough states have been generated.

An alternative approach, developed by Yang and Kubat (1989), generates states by traversing a certain balanced, rooted tree, T, of 'partial' states. The nodes of T at level i are at distance i from the root and correspond to states in which the modes of components $k \le i$ have been fully specified but the remaining modes have not. Each node at level i has m_{i+1} successors corresponding to the possible modes for component $i+1$. The leaves of the tree (at level n) correspond to the states X of the system. The algorithm proceeds by generating only those portions of the tree T that are of interest in searching for a leaf of maximum probability. A dynamic-programming approach, aided by the ordering of modes in (8.7), allows such a leaf to be determined in one transversal from the root to the leaf. The return traversal from this leaf back to the root then updates the necessary pointers to enable the next most probable state to be found subsequently.

A third algorithm, which has proved quite effective in practice, is based on a method proposed by Gaebler and Chen (1987). It also represents a natural extension of the algorithm GENERATE discussed in Section 8.2, and so we present it here in some detail. In addition to assuming that (8.7) holds for each component, we also suppose that the components have been renumbered so that

$$p_{i1}/p_{i0} \ge p_{k1}/p_{k0} \quad \text{for } i < k. \tag{8.8}$$

This numbering of components generalizes the ordering (8.2) imposed in the case of binary systems. In view of this numbering, $X_0 = (0, 0, \ldots, 0)$ is the most probable state and $X_1 = (1, 0, \ldots, 0)$ is the second most probable state. A set of candidate states $\mathscr{C}$, initialized to contain X_1, is maintained at each iteration. Then a state X of maximum probability is removed from $\mathscr{C}$, and this state is used to generate at most three successor states (all of no larger probability than X) which are added to $\mathscr{C}$.

To describe the rules for generating successor states in this algorithm, it is first convenient to define the function F on states $X \ne X_0$ by $F(X) = \min\{i: X[i] \ne 0\}$. That is, $F(X)$ is the index of the lowest-numbered component of X not in mode 0. Also, define a mode replacement operator R_{ij}, which, when applied to state X, changes $X[i]$ to mode j. Notice that, if $X[i] = k$ then $p(R_{ij}X) = p(X) \cdot p_{ij}/p_{ik}$. Now suppose X is the state just removed from $\mathscr{C}$, and let $f = F(X)$, $j = X[f]$, $m = m_f$. Then the following successor states are generated from X:

(1) if $j < m - 1$ then $R_{f, j+1}X$ is a successor;

(2) if $j = 1$, $f < n$, and $X[f + 1] = 0$ then $R_{f, 0}R_{f+1, 1}X$ is a successor;

(3) if $f > 1$, then $R_{11}X$ is a successor.

For example, state $X = (0, 1, 0, 2, 1, 0)$ has successor states $X^{(1)} = (0, 2, 0, 2, 1, 0)$, $X^{(2)} = (0, 0, 1, 2, 1, 0)$, and $X^{(3)} = (1, 1, 0, 2, 1, 0)$, respectively. For completeness, we consider that state X_0 has the single successor state X_1.

To verify the correctness of this procedure, we first notice from (8.7)–(8.8) that each successor of X has probability no larger than that of X. Thus, selecting the maximum probability state from $\mathscr{C}$ ensures that the algorithm will produce states in order of nonincreasing probability. It remains to show that every state $Y \neq X_0$ will eventually be generated by this procedure if run to completion. First, it follows from the rules given above that each state $Y \neq X_0$ has a unique predecessor state X (that generated it). Also, this predecessor state X is lexicographically smaller than Y, when the components are read from right to left. Hence, iterating the predecessor mapping will eventually lead to state X_1, and so every state in $\mathscr{S}$ can be generated.

To illustrate this method for generating states, suppose that there are four components each assuming three modes, with corresponding probabilities given in Table 8.3. The modes have been numbered so that mode probabilities are nonincreasing for each component, and components have been ordered by decreasing values of p_{i1}/p_{i0}. Table 8.4 shows, for the first 10 steps of the multistate algorithm, the state generated at each step, as well as the candidate set after the most probable state has been removed. In this example, the first 10 states account for over 80% of the total probability, and only 7 more states are needed to achieve 90% coverage. In addition, the size of the candidate set $\mathscr{C}$ remains quite modest throughout this particular example. More extensive computational results with this method (Shier et al. 1990) confirm that this algorithm is quite effective in generating most probable states of multistate systems.

Table 8.3 Probabilities for a four-component three-mode example

Component	Mode		
	0	1	2
1	0.70	0.20	0.10
2	0.70	0.15	0.15
3	0.80	0.15	0.05
4	0.90	0.05	0.05

Table 8.4 Generated states and candidate sets for the multistate algorithm

Generated state	Cumulative probability	Candidate set
(0, 0, 0, 0)	0.3528	(1, 0, 0, 0)
(1, 0, 0, 0)	0.4536	(2, 0, 0, 0), (0, 1, 0, 0)
(0, 1, 0, 0)	0.5292	(2, 0, 0, 0), (0, 2, 0, 0), (0, 0, 1, 0), (1, 1, 0, 0)
(0, 2, 0, 0)	0.6048	(2, 0, 0, 0), (0, 0, 1, 0), (1, 1, 0, 0), (1, 2, 0, 0)
(0, 0, 1, 0)	0.6710	(2, 0, 0, 0), (1, 1, 0, 0), (1, 2, 0, 0), (0, 0, 2, 0), (0, 0, 0, 1), (1, 0, 1, 0)
(2, 0, 0, 0)	0.7214	(1, 1, 0, 0), (1, 2, 0, 0), (0, 0, 2, 0), (0, 0, 0, 1), (1, 0, 1, 0)
(0, 0, 2, 0)	0.7434	(1, 1, 0, 0), (1, 2, 0, 0), (0, 0, 0, 1), (1, 0, 1, 0), (1, 0, 2, 0)
(1, 1, 0, 0)	0.7650	(1, 2, 0, 0), (0, 0, 0, 1), (1, 0, 1, 0), (1, 0, 2, 0), (2, 1, 0, 0)
(1, 2, 0, 0)	0.7866	(0, 0, 0, 1), (1, 0, 1, 0), (1, 0, 2, 0), (2, 1, 0, 0), (2, 2, 0, 0)
(0, 0, 0, 1)	0.8062	(1, 0, 1, 0), (1, 0, 2, 0), (2, 1, 0, 0), (2, 2, 0, 0), (0, 0, 0, 2), (1, 0, 0, 1)

8.5 Chapter notes

Li and Silvester (1984) introduced the idea of generating the most probable states in order to approximate the performance of computer networks. Their algorithm was subsequently improved by Lam and Li (1986a) and by Shier (1988b). The lattice-based approach developed in Section 8.1 can be found in Valvo et al. (1987), which also discusses computational experience with the resulting algorithm (Section 8.2).

Chiou and Li (1986) first proposed an algorithm for enumerating, in order, the states of a multistate system. Gaebler and Chen (1987) and Yang and Kubat (1989) subsequently developed improved algorithms for carrying out the generation of most probable states. Further computational refinements are reported by Shier et al. (1990) and by Yang (1989). In addition, Shier et al. (1990) present an empirical comparison of improved versions of three different approaches for the multistate problem; the procedure described in Section 8.4, which extends algorithm GENERATE of Section 8.2, exhibited the best computational performance of all the methods tested.

9

Stochastic shortest-path problems

The final section of Chapter 8 introduced multistate systems: those in which the components assume a finite number of distinct modes. One particular instance of this type of system frequently arises in the modelling of various transportation, communication, and project planning networks. For example, it may be required to transport material from a warehouse to a distribution centre. If the travel times along the associated highway network are known, then it would be desirable to ship the material along a timewise shortest path between the warehouse and the distribution centre. In reality, the travel times will be subject to uncertainty, depending, for instance, on the volume of traffic, the timing of traffic lights, and construction delays. Consequently, each road segment in the highway network can be viewed as having a range of travel times, specified by a probability distribution. Another example concerns the scheduling of complex projects, which can be modelled using a network of tasks (corresponding to edges) and their precedences (Elmaghraby 1977). If the duration of each task is known exactly, then the completion time in such a project planning network corresponds to a longest (or 'critical') path from the project start node to the project completion node. Most realistically, however, the actual duration of each task is subject to uncertainty and so it is best represented by a probability distribution.

In both of these instances, the underlying network can be viewed as a multistate system in which edges assume various lengths with certain probabilities. Since the optimal path will generally change as the state of the system changes, it is not meaningful to talk about *the* best path in the network. Rather it is more informative to describe the stochastic properties of the optimal path *length* in the system. Accordingly, this chapter studies the distribution of the shortest (or longest) path length in a network with stochastically varying edge lengths. Of course, once this probability distribution has been determined, the expected path length (as well as its variance) can be easily calculated.

The methodology of Section 8.4 can certainly be applied to approximate the distribution of optimal path length in such a stochastic network, by generating high-probability states and evaluating the optimal path in each such state. The emphasis of this chapter, however, is to develop approaches for finding the exact path length distribution. We will identify an underlying algebraic structure that allows further insight into the nature of this problem and that reveals its similarity to algebraic problems studied in earlier chapters. An alternative approach is also studied that exploits certain topological structures present in the network to simplify the calculation of the optimal path length distribution. This latter method employs a special type of 'factoring,' which allows a stochastic network to be decomposed into an equivalent set of smaller, generally less complex, subnetworks. Computational evidence shows that this structural factoring approach significantly reduces the computational effort required to solve a problem, especially when compared to complete enumeration.

9.1 Algebraic structure

Let $G = (N, E)$ denote a directed network having two distinguished nodes s, t. Each edge $i \in E$ assumes a finite number of lengths $\{l_{ir}\}$ with corresponding probabilities $\{p_{ir}\}$. The length of edge i is thus a random variable X_i with probability distribution given by $\Pr(X_i = l_{ir}) = p_{ir}$. It is also supposed that the random variables for distinct edges are statistically independent. For each state of the network, specified by the particular lengths assumed by each edge, there is a shortest s-t path and its length is denoted X. The *stochastic shortest-path problem* studied here is to determine the distribution of the random variable X from the given random variables X_i and the structure of the network G.

A special case of this problem is, in fact, the two-terminal reliability problem, defined relative to some given network H in which edge k has probability p_k of functioning, independent of all other edges. From H, construct the stochastic network G involving the same nodes and edges as H. Each edge k in G will assume the lengths 0 and 1, with probabilities p_k and $1 - p_k$, respectively. If the shortest s-t path length in G is 0, then there must be some path in G along which all edge lengths are 0, corresponding to a path in H composed of functioning edges. Thus the probability that the shortest s-t path in G has length 0 is precisely the two-terminal reliability of the original network H. Since this latter problem is #P-hard, the stochastic shortest-path problem is computationally difficult as well, and efficient algorithms are unlikely to be found.

An essential feature of the stochastic shortest-path problem, in contrast to the efficiently solvable shortest-path problem for deterministic networks, is the statistical dependence created when different paths share common edges. The algebraic structure to be defined in this section thus needs to explicitly account for the identity of individual edges. To this end, we use the binary variable x_{ir} to indicate that edge i assumes its rth length l_{ir}. Since edge i has, at any instant, precisely one length, then

$\sum_r x_{ir} = 1$ must hold. For notational simplicity, it will be convenient to assume that each length l_{ir} is a nonnegative integer. Then the *generating function* $f(z)$ for edge i is defined by the following polynomial in the edge variables and the indeterminate z:

$$f(z) = f(i; z) = \sum_r x_{ir} z^{l_{ir}}. \tag{9.1}$$

Moreover, we define the multiplication of edge variables using

$$x_{ir} \odot x_{ir} = x_{ir} \tag{9.2}$$

$$x_{ir} \odot x_{iw} = 0 \quad \text{for } r \neq w \tag{9.3}$$

$$x_{ir} \odot x_{jw} = x_{ir} x_{jw} \quad \text{for } i \neq j. \tag{9.4}$$

The above relations simply formalize the important characteristics of the binary edge variables. This multiplication is extended by associativity in the natural way to products of terms: e.g., $(x_{12} x_{23} x_{51}) \odot (x_{23} x_{41}) = x_{12} x_{23} x_{41} x_{51}$ and $(x_{12} x_{23} x_{51}) \odot (x_{22} x_{41}) = 0$. Now suppose that $f = \sum f_j z^j$ and $g = \sum g_k z^k$ are the generating functions (9.1) for two edges. We define the following two operations on such functions:

$$h = f \otimes g \iff h_r = \sum_{j+k=r} (f_j \odot g_k) \tag{9.5}$$

$$h = f \oplus g \iff h_r = \sum_{\min(j,k)=r} (f_j \odot g_k). \tag{9.6}$$

For example, if $f = x_{11} x_{21} z^2 + x_{12} x_{31} z^3$ and $g = x_{21} z^2 + x_{22} x_{31} z^4$ then, using (9.2)–(9.4), we obtain

$$f \otimes g = x_{11} x_{21} z^4 + x_{12} x_{21} x_{31} z^5 + x_{12} x_{22} x_{31} z^7,$$

$$f \oplus g = (x_{11} x_{21} + x_{12} x_{21} x_{31}) z^2 + x_{12} x_{22} x_{31} z^3.$$

The motivation for these operations is again so that $\oplus$ is appropriate for combining two edges in parallel while $\otimes$ is appropriate for combining two edges in series. Let $\mathscr{S}$ denote the set of all functions that can be obtained from the edge generating functions (9.1) by finite application of (9.5)–(9.6). Then the following properties hold for $f, g, h \in \mathscr{S}$:

$$f \oplus f = f$$

$$f \oplus g = g \oplus f \qquad f \otimes g = g \otimes f$$

$$f \oplus (g \oplus h) = (f \oplus g) \oplus h \qquad f \otimes (g \otimes h) = (f \otimes g) \otimes h$$

$$f \oplus (f \otimes g) = f$$

$$f \otimes (g \oplus h) = (f \otimes g) \oplus (f \otimes h)$$

$$f \oplus 0 = f \qquad f \otimes 1 = f.$$

These properties are similar to those encountered in (3.1)–(3.6) with regard to the algebraic structure presented for the two-terminal reliability problem. Moreover, it is not difficult to see that an expression for the distribution of shortest path lengths can be succinctly written in terms of these two operations. Namely, let $\mathcal{P}_{st}$ denote the set of all simple s-t paths P in G. Also, define the path value $v(P)$ for $P \in \mathcal{P}_{st}$ by the product (9.5) of the edge generating functions along path P. Then

$$\bigoplus \sum_{P \in \mathcal{P}_{st}} v(P)$$

provides a symbolic form for the shortest path length distribution. That is to say, substitution of the values p_{ir} for x_{ir} into this function yields a generating function $f^*(z)$ for the distribution of shortest path lengths in the network.

The similarity of this algebraic structure to the structures encountered earlier suggests defining an order relation $\leqslant$ on $f, g \in \mathcal{S}$ by:

$$f \leqslant g \Leftrightarrow f \oplus g = g. \tag{9.7}$$

From the above properties of $\oplus$ it follows that $\leqslant$ is, in fact, a partial order. A useful interpretation can be given to (9.7) by noting that

$$\mu_f = \frac{df}{dz}(z)\bigg|_{z=1}$$

is the average value of the distribution represented by the generating function f. (Here it is assumed that the probabilities p_{ir} have been substituted for the corresponding edge variables x_{ir}.) From the definition of $\oplus$ in (9.6), it can be readily shown that $\mu_{f \oplus g} \leqslant \mu_f$. Therefore, $f \leqslant g$ implies that $\mu_f \geqslant \mu_g$: i.e., smaller distributions in the partial order have larger means.

It is also possible to adapt various algebraic solution methods, developed for network reliability problems, to the solution of stochastic shortest-path problems. For example, the necessary properties hold for the iterative methods of Chapter 3 to be valid in the present context. Specifically, the list-directed scheme of Section 3.4 can be applied: node labels $L(w)$ are maintained and updated at each step using $L(w) := L(w) \oplus [L(v) \otimes f]$, where f is the generating function for edge (v, w). Implementation of this scheme then yields a sequence of approximating distributions for the stochastic shortest-path problem. The resulting approximations are guaranteed to be nondecreasing in terms of the partial order $\leqslant$ in (9.7). By our previous observation, the expected values of successive distributions will thus be nonincreasing.

9.2 Fundamental reductions

The preceding section has indicated how the stochastic shortest-path problem can be investigated using a symbolic method. In the remaining sections we explore an alternative approach that works directly with the numerical quantities p_{ir}, rather than with the edge variables x_{ir} and edge generating functions $f(z)$. Our starting point will be the observation that certain fundamental reductions (e.g., series and parallel reductions) can be directly invoked to simplify a given network. This suggests identifying and exploiting other special structures that may be present in the network.

For each edge i of the given network $G = (N, E)$ we denote by A_i the set of possible (integer) lengths assumed by that edge. Recall that the random variable X_i for edge i has the known probability distribution $\Pr(X_i = l_{ir})$, for $l_{ir} \in A_i$. If i and k are a pair of series edges then they can be replaced by a single edge m having the edge length set

$$A_i \otimes A_k = \{u + v : u \in A_i, v \in A_k\}.$$

By the assumed independence of the random variables X_i and X_k, the probability of any derived length $w \in A_m$ is given by the discrete convolution of the constituent edge length distributions:

$$\Pr(X_m = w) = \sum_{u+v=w} \Pr(X_i = u) \Pr(X_k = v). \tag{9.8}$$

On the other hand, if two edges i and k occur in parallel with corresponding edge length sets A_i and A_k, then they can be replaced by a single edge m with edge length set

$$A_i \oplus A_k = \{\min(u, v) : u \in A_i, v \in A_k\}.$$

Again, by independence, the probability of any length $w \in A_m$ is given by

$$\Pr(X_m = w) = \sum_{\min(u,v)=w} \Pr(X_i = u) \Pr(X_k = v). \tag{9.9}$$

These formulas represent appropriate simplifications of eqns (9.5)–(9.6), under the crucial assumption of edge independence. If the network G can be reduced to a single edge (s, t) by a succession of these fundamental series and parallel reductions, then the shortest path length distribution can be easily obtained by applying the above formulas (Martin 1965). Consider, for example, the series–parallel network G shown in Fig. 9.1(a), with each edge having the equiprobable lengths $\{1, 2\}$. The network can be reduced using (9.8)–(9.9) to the two-node network shown in Fig. 9.1(b). In this example the shortest path length in G assumes values $\{3, 4, 5, 6\}$ with the indicated probabilities.

Fundamental reductions

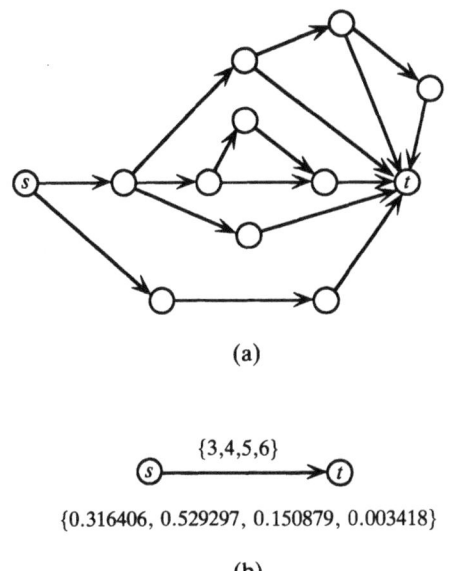

Fig. 9.1 A series–parallel network and its complete reduction.

A further fundamental reduction (called the *figure-eight* reduction) can be applied to certain directed cycles, as indicated in Fig. 9.2. (Notice the similarity of this reduction to the two-neighbour reduction shown in Fig. 2.3(c).) The central node j of this configuration is required to have indegree 2 and outdegree 2. The two replacement edges have edge length sets $A_1 \otimes A_3$ and $A_2 \otimes A_4$; the associated probabilities are obtained by convoluting the appropriate two edge distributions via (9.8). The validity of this reduction rule follows from the fact that in any particular realization the shortest path will never include edges in both A_1 and A_2, nor will it include edges in both A_3 and A_4. Notice that the two edges created will maintain the independence required for further processing.

The advantage of using these three fundamental reductions is that they yield a network possessing the same shortest-path distribution. Moreover, this equivalent network has a smaller number of edges and possibly fewer

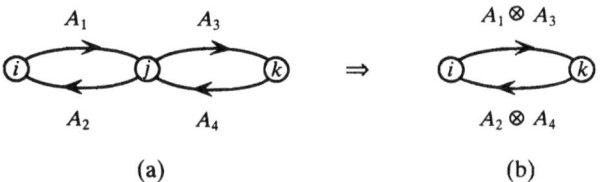

Fig. 9.2 A figure-eight structure and its reduction.

nodes. Such reductions have been previously discussed in regard to network reliability calculations in Section 2.4, and they have been found to be quite useful in the practical solution of reliability problems (Ball and Cameron 1986; Page and Perry 1988, 1989; Resende 1986). Given the success of this approach, we are encouraged to investigate other structures that can be used for simplifying a given network.

9.3 Conditioned reductions

The conditioned reductions described next are performed by identifying specific topological structures centred around a particular node c of the structure. Our goal is to remove node c from the network and replace the various incident edges (i, c) and (c, j) by new edges (i, j). This process needs to be done carefully to ensure that the resulting edge length variables X_m remain independent. To the extent that this procedure can be successfully carried out, it affords a substantial reduction in computational effort compared to total state enumeration relative to all edges incident with c. Because the focus is now on edges incident with node c, the various edge length sets and random variables A_m, X_m will be subscripted by the ordered pair of nodes defining edge m.

Recall that, in a series structure, some node $c \neq s, t$ has indegree 1 (one entering edge) and outdegree 1 (one leaving edge). A direct extension of this case is a *fan* structure, in which node $c \neq s, t$ has indegree 1 and outdegree > 1. Clearly there is an analogous case when c has outdegree 1 and indegree > 1. The former case is illustrated in Fig. 9.3(a). If node c were removed from Fig. 9.3(a) and the new edge length sets $A_{ij} = A_{ic} \otimes A_{cj}$ were formed, then the derived random variables X_{ij} would be dependent. Instead, we identify edge (i, c) as a *factoring edge*, and for each $u_{ic} \in A_{ic}$ we create a subproblem S_{ic} in which edge (i, c) has the deterministic length u_{ic}. There are $|A_{ic}|$ such subproblems and the probability associated with each subproblem is $\Pr(X_{ic} = u_{ic})$. In subproblem S_{ic} we can reduce the fan structure by convoluting via (9.8) the degenerate distribution for $\{u_{ic}\}$ with the distribution for each A_{cj}, as

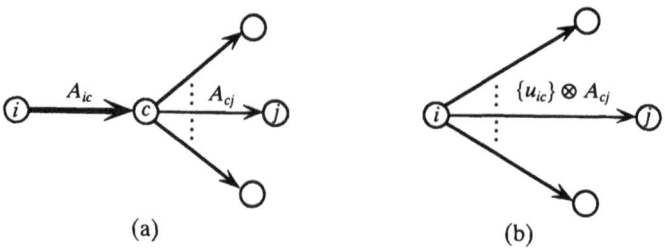

Fig. 9.3 Reduction of a fan structure.

indicated in Fig. 9.3(b). The new edge random variables X_{ij} produced in this way are now independent since, in any realization, a shortest path will use at most one of the edges leaving node c.

The next three constructs focus on situations in which the central node $c \neq s, t$ has indegree 2 or outdegree 2. Fig. 9.4(a) describes a *loop* construct. The possible subpaths involving c that might be considered in a shortest path are then (i, c, k), (k, c, j), and (i, c, j). We would like to remove node c from this loop construct and obtain the three replacement edges (i, k), (k, j), and (i, j). However, since edges (i, c) and (c, j) are used more than once in defining these new edges, the random variables X_{ik}, X_{kj}, X_{ij} will be correlated. To ensure independence of these random variables, we specify edges (i, c) and (c, j) to be factoring edges: namely, for each $u_{ic} \in A_{ic}$ and $u_{cj} \in A_{cj}$ we form the subproblem $S_{ic,cj}$ in which edges (i, c) and (c, j) have these given deterministic lengths. In each such subproblem, node c can be eliminated, resulting in the subnetwork depicted in Fig. 9.4(b). The associated edge distributions are again found by discrete convolution, based on the constituent edge sets indicated by Fig. 9.4(b). Notice that two factoring edges, shown in bold in Fig. 9.4(a), are required and the number of subproblems produced is $|A_{ic}| |A_{cj}|$. The probability $\Pr(X_{ic} = u_{ic}) \Pr(X_{cj} = u_{cj})$ is associated with each generated subproblem $S_{ic,cj}$.

Two other constructs have proved useful in reducing directed networks. The first, in which c has outdegree 2 and indegree > 2, extends the loop construct. Removal of the central node $c \neq s, t$ in Fig. 9.5(a) produces the topology shown in Fig. 9.5(b). All edges indicated in bold in Fig. 9.5(a) must now be used as factoring edges; each combination of lengths chosen from factoring edges generates a new subproblem to be solved. In the second construct, node c also has outdegree 2 and indegree > 2, as illustrated in Fig. 9.6(a); the associated factoring edges are shown in bold. Fig. 9.6(b) describes the topological form of each subproblem obtained by conditioning on the lengths of the factoring edges and then removing node c. Similar constructs apply if the edges of these figures are reversed in orientation.

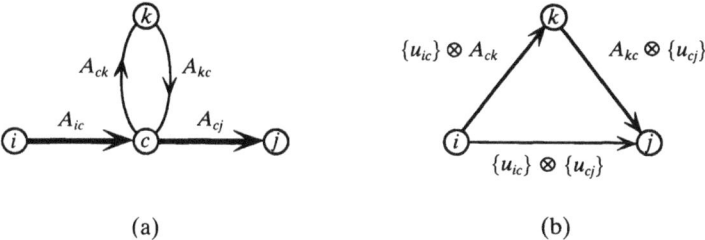

(a)   (b)

Fig. 9.4 Reduction of a loop construct.

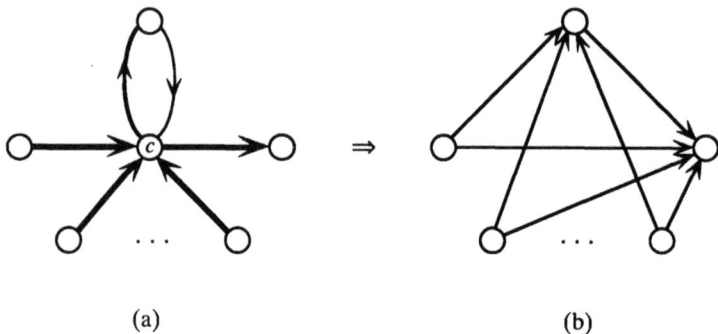

Fig. 9.5 A generalized loop construct.

These particular constructs were investigated because they represent simple logical extensions of the series reduction. They are also present in a variety of sparse networks encountered in practice. Unlike the fundamental reductions, these conditioned reductions generate a sequence of subproblems that must then be solved. In each case the number of generated subproblems is smaller than the number of subproblems obtained if total enumeration were applied to all edges incident with the central node c. Notice that the generated subproblems can in turn be recursively processed by the various fundamental and conditioned reductions. The next section discusses the implementation of this structural factoring approach and illustrates the approach on selected examples.

9.4 Computational results

In this section the strategy of successively applying fundamental and conditioned reductions will be illustrated by several examples. While the

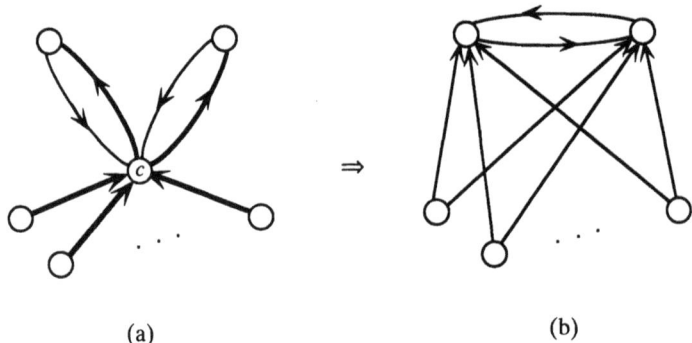

Fig. 9.6 A generalized double-loop construct.

computational effort required by this approach will grow relentlessly with the problem size, the structural factoring approach has enabled some fairly complex problems to be solved exactly. Indeed, the ability to solve certain nontrivial problems exactly can serve as a useful baseline for assessing the accuracy of various approximation schemes necessary to solve large problems.

It is convenient to visualize the reduction algorithm by a computation tree whose root node represents the given network G. When a conditioned reduction decomposes a subproblem S into an equivalent set of subproblems $\{S_k\}$, the subproblems S_k are placed as children of node S. If none of the four conditioned reductions can be applied to a given node, then the corresponding problem generates an equivalent set of subproblems by (total) factoring on all edges incident with the selected central node c. The leaves of this tree correspond to subproblems that can be completely reduced (to a single edge) by applying the three fundamental reductions: series, parallel, and figure-eight. The shortest path length distribution for G is obtained by weighting the optimal path length distributions found at the leaves by the appropriate conditional probabilities associated with these subproblems.

In an attempt to minimize the width of the computation tree, the node chosen for conditioned reductions (or total factoring) was selected to have the minimum possible indegree or outdegree. Ties were broken on the basis of the smallest number of generated subproblems at the next level. Conceptually, a depth-first search of the computation tree is carried out, so that only a small number of subproblems are stored at any one time. If there are $n = |N|$ nodes in G, then this computation tree has depth at most $n - 3$. A program to carry out the reduction process was implemented in FORTRAN 77 on a Microvax 3200 computer (Hayhurst 1989). This program was used to find the exact distribution of the shortest and longest path length for several standard networks in the literature.

The first test example is the 'crossing' network with 6 nodes and 8 edges studied by Kleindorfer (1971). Each edge assumes a length in the set $\{1, 2, 3, 4, 5\}$ with equal probability. The computation tree for this acyclic network consists of three levels (depth 2), corresponding to two nested applications of the fan reduction. The entire procedure analysed 31 subproblems to calculate the distribution of the critical (longest) path length, requiring a total computation time of 0.18 seconds. Complete enumeration would have required evaluating $5^8 = 390\,625$ possible states to determine the exact distribution. The true mean completion time for this project network is 11.20, which can be compared to the lower and upper bounds of 9.00 and 11.36 obtained by Kleindorfer (1971), and the bounds 10.60 and 11.36 subsequently obtained by Shogan (1977b).

More generally, any two-terminal acyclic network can be completely simplified to a single edge using only series, parallel, and fan reductions.

This follows since, if no series or parallel reductions are applicable, we can select as node c the node adjacent to the source s having minimum topological number. In view of this choice, node c has only one entering edge (s, c) and so the fan reduction can be carried out, yielding a new acyclic network. Continuing this process shows that at most $n - 3$ such fan reductions are necessary for complete reduction to an edge. In fact, far fewer such reductions are needed in practice.

The second example is the cyclic network on 7 nodes and 15 edges studied by Shogan (1976). For this example the set of lengths for each edge is $\{0, 2, 4\}$ with corresponding probabilities $\{0.4, 0.2, 0.4\}$. The shortest path length distribution, shown in Table 9.1, was obtained in 4.5 seconds and required 1 528 subproblems to be analysed. Only loop and fan reductions were necessary to completely reduce the network. Furthermore the maximum number of subproblems that required storage at any one time was 27. This problem can also be considered as defining a two-terminal reliability problem with each $p_k = \Pr(X_k = 0) = 0.4$. The exact two-terminal reliability value 0.30475 is found in Table 9.1 by consulting the computed probability that the shortest path has length 0. Shogan (1976) gives a lower bound of 0.187 and an upper bound of 0.350 on the two-terminal reliability for this network.

The final example is an acyclic network on 10 nodes and 16 edges studied by Fulkerson (1962). The shortest path length distribution was calculated when each edge had its length selected from $\{0, 1\}$ with equal probability. This generated 127 subproblems and required a total of 0.34 seconds of computation time. A more complex problem was solved by allowing each edge to assume values from $\{1, 2, 3, 4\}$ with equal probability. This generated 5 461 subproblems and required 26.8 seconds; a maximum of 19 subproblems had to be stored at any one time. By contrast, complete enumeration of the $4^{16} \cong 4.3 \times 10^9$ states of this latter network would have been totally out of the question.

Table 9.1 Distribution of shortest path length for Shogan example

Length	Probability
0	0.30475
2	0.28108
4	0.27844
6	0.09870
8	0.03392
10	0.00285
12	0.00026

9.5 Chapter notes

The stochastic shortest-path problem is #P-hard, even for acyclic networks. In fact, just computing the expected value of the shortest (or longest) path length in such networks is provably difficult (Hagstrom 1988). Mirchandani (1976) discusses the problem of determining the expected shortest path length in a stochastic network and shows how to transform this into a problem on an 'emergency equivalent' network. Both the reliability and expected path length in this latter network can be found simultaneously using a recursive algorithm. Hagstrom (1990) has used a factoring approach, together with the topological numbering of nodes in an acyclic network, to compute the exact distribution of longest path length in a PERT network. The particular factoring approach presented in this chapter can be found in Hayhurst and Shier (1990). For the case of acyclic networks (which can be reduced using series, parallel, and fan constructs), Elmaghraby *et al.* (1989) apply the factoring approach for obtaining shortest or longest path length distributions. In addition, they discuss an 'optimal' node selection strategy for minimizing the number of fan reductions that need to be invoked.

A number of methods have been proposed for approximating the mean completion time in PERT networks (Elmaghraby 1967; Fulkerson 1962; Malcolm *et al.* 1959). Dodin (1985), Kleindorfer (1971), and Shogan (1977b) provide, instead, approximations to the distribution of the critical path length. Alternatively, Monte Carlo methods have been employed to approximate the PERT distribution (Burt and Garman 1971; Van Slyke 1963). An excellent summary of both exact and approximate methods for analysing PERT networks is provided in the book of Elmaghraby (1977).

Glossary of terms

acyclic: a directed network that does not contain any directed cycles; an undirected network that does not contain any cycles.

adjacent: nodes i and j that are joined by an edge.

antichain: a set of elements in a partially ordered set with any two such elements being incomparable.

arc: a directed edge.

bipartite: a network whose node set can be partitioned into two sets X and Y, such that each edge joins a node in X to a node in Y.

chain: a set of elements in a partially ordered set with any two such elements being comparable.

comparable: two elements x, y of a partially ordered set such that either $x \leqslant y$ or $y \leqslant x$ holds.

coherent system: a system of components each of which is subject to failure; the system must be monotone with respect to repair of failed components.

connected: an undirected network in which every pair of distinct nodes is joined by a path.

connected component: in a given undirected network, a maximal subnetwork that is connected.

convex: a set C of elements in a partially ordered set such that $x, y \in C$ and $x \leqslant z \leqslant y$ imply that $z \in C$.

cut: the set of all edges (i, j) joining nodes i in some set X to nodes j not in X

cut node: a node disconnecting set of size one.

cutset: a minimal edge disconnecting set.

cycle: a path from a node to itself.

degree: the number of edges incident with a node in an undirected network.

depth: in a tree, the maximum number of edges in a path joining any node to the root.

diameter: the maximum distance between nodes in a network.

distance: the minimum length of a path between two nodes in a network.

edge: an ordered pair (i, j) of nodes in the case of a directed network, or an unordered pair $[i, j]$ of nodes in the case of an undirected network.

edge cover: a set of edges that is incident with every node of a given network.

edge connectivity: the minimum size of an edge disconnecting set.

edge disconnecting set: any set of edges in a connected network whose removal disconnects the network.

failure set: a set of components in a coherent system whose failure renders the system inoperable.

graph: a set of nodes together with a set of edges joining those nodes.

Hamiltonian cycle: a cycle that includes every node of a network exactly once.

Hasse diagram: representation of a partially ordered set in which distinct elements are represented by distinct points. If x and y are distinct comparable elements with $x \geqslant y$, then x and y are joined by a line segment descending from x to y, but those relations implied by transitivity are not explicitly drawn.

incident: describing the relationship between an edge (i, j) and either of its nodes (i or j).

incomparable: two elements x, y of a partially ordered set such that neither $x \leqslant y$ nor $y \leqslant x$ holds.

indegree: the number of directed edges entering a node.

induced subnetwork: in a given network, a subset of nodes together with all edges joining them in the original network.

irrelevant: an edge not on some simple s-t path.

lattice: a partially ordered set in which any two elements x, y possess both a greatest lower bound $x \wedge y$ and a least upper bound $x \vee y$.

leaf: a node of a tree at maximal distance from the root.

length: the number of edges in a path.

matching: a set of edges in a network, no two of which are incident with the same node.

mincut: a minimal failure set in a coherent system.

minpath: a minimal success set in a coherent system.

network: a set of nodes together with a set of edges joining them; frequently there is also quantitative information associated with the edges of a network.

node connectivity: the minimum size of a node disconnecting set.

node disconnecting set: any set of nodes in a connected network whose removal either disconnects G or produces a single-node network.

outdegree: the number of directed edges leaving a node.

parallel edges: edges that extend from the same initial node to the same final node.

partial order: a binary relation that is reflexive, antisymmetric, and transitive.

path: a sequence of nodes $i_0, i_1, \ldots, i_k$ with each (i_{r-1}, i_r) being an edge of the network.

perfect matching: a matching that is also an edge cover.

planar: a network that can be drawn in the plane in such a way that edges meet one another only at nodes.

ranked: a partially ordered set which can be partitioned into linearly ordered sets of elements so that lines of the Hasse diagram join elements only in successive sets.

relevant: an edge that occurs on some simple s-t path.

relevant subnetwork: a subnetwork each of whose edges is relevant.

self loop: an edge joining a node to itself.

semilattice: a (meet) semilattice is a partially ordered set in which any two elements x, y possess a greatest lower bound $x \wedge y$.

series edges: two edges (i, j) and (j, k) in a network that are the only edges incident with node j.

simple path: a path in which all nodes (except possibly the first and last) are distinct.

subnetwork: in a given network, a subset of nodes together with certain edges joining them in the original network.

subtree: a subnetwork that forms a tree.

success set: a set of components in a coherent system whose operation ensures that the system operates.

topological numbering: a numbering of nodes in a directed (acyclic) network such that edges only extend from lower numbered nodes to higher numbered nodes.

tree: an undirected network that is acyclic and connected; a directed network whose underlying undirected network forms a tree.

Vandermonde: a square $n \times n$ matrix $V = (v_{ij})$ with $v_{ij} = r_i^{j-1}$ for distinct values $r_1, \ldots, r_n$.

References

Abel, U. and Bicker, R. (1982). Determination of all minimal cut-sets between a vertex pair in an undirected graph. *IEEE Transactions on Reliability*, **R-31**, 167–71.

AboElFotoh, H. M. and Colbourn, C. J. (1989). Series–parallel bounds for the two-terminal reliability problem. *ORSA Journal on Computing*, **1**, 209–22.

Agrawal, A. and Barlow, R. E. (1984). A survey of network reliability and domination theory. *Operations Research*, **32**, 478–92.

Aho, A. V., Hopcroft, J. E., and Ullman, J. D. (1974). *The design and analysis of computer algorithms*. Addison-Wesley, Reading.

Amin, A. T., Siegrist, K. T., and Slater, P. J. (1988). Pair-connected reliability of communication networks with vertex failures. *Congressus Numerantium*, **67**, 233–42.

Assous, J. Y. (1986). First- and second-order bounds on terminal reliability. *Networks*, **16**, 319–29.

Backhouse, R. C. and Carré, B. A. (1975). Regular algebra applied to path-finding problems. *Journal of the Institute of Mathematics and its Applications*, **15**, 161–86.

Bagga, K. S., Beineke, L. W., Lipman, M. J., and Pippert, R. E. (1989). A survey of integrity. Technical Report. Department of Mathematical Sciences, Indiana University–Purdue University at Fort Wayne.

Ball, M. O. (1980). Complexity of network reliability computations. *Networks*, **10**, 153–65.

Ball, M. O. (1986). Computational complexity of network reliability analysis: an overview. *IEEE Transactions on Reliability*, **R-35**, 230–9.

Ball, M. O. and Cameron, E. P. (1986). Experiments with network reliability analysis algorithms. In *Proceedings of the 17th annual conference on modeling and simulation* (ed. W. G. Vogt and M. H. Mickle), pp. 1799–803, Pittsburgh.

Ball, M. O. and Provan, J. S. (1982). Bounds on the reliability polynomial for shellable independence systems. *SIAM Journal on Algebraic and Discrete Methods*, **3**, 166–81.

Ball, M. O. and Provan, J. S. (1983). Calculating bounds on reachability and connectedness in stochastic networks. *Networks*, **13**, 253–78.

Ball, M. O. and Provan, J. S. (1987). Computing K-terminal reliability in time polynomial in the number of (s, K)-quasicuts. In *Fourth Army conference on applied mathematics and computing*, pp. 901–7. Army Research Office, Washington.

Ball, M. O. and Provan, J. S. (1988). Disjoint products and efficient computation of reliability. *Operations Research*, **36**, 703–15.

Ball, M. O., Provan, J. S., and Shier, D. R. (1989). Reliability covering problems. Technical Report 89-02. Department of Mathematics, College of William and Mary.

Barefoot, C. A., Entringer, R., and Swart, H. (1987). Vulnerability in graphs—a comparative survey. *Journal of Combinatorial Mathematics and Combinatorial Computing*, **1**, 12–22.

Barlow, R. E. (1984). Mathematical theory of reliability: a historical perspective. *IEEE Transactions on Reliability*, **R-33**, 16–20.

Barlow, R. E. and Proschan, F. (1981). *Statistical theory of reliability and life testing*. To Begin With, Silver Spring.

Barlow, R. E. and Singpurwalla, N. D. (1985). Assessing the reliability of computer software and computer networks: an opportunity for partnership with computer scientists. *The American Statistician*, **39**, 88–94.

Beineke, L. W. (1989). Explorations into graph vulnerability. Technical Report. Department of Mathematical Sciences, Indiana University–Purdue University at Fort Wayne.

Beineke, L. W. and Harary, F. (1967). The connectivity function of a graph. *Mathematika*, **14**, 197–202.

Beineke, L. W., Lesniak, L., and Pippert, R. E. (1990). On the edge-separation sequence of a graph. Technical Report. Department of Mathematical Sciences, Indiana University–Purdue University at Fort Wayne.

Bellmore, M. and Jensen, P. A. (1970). An implicit enumeration scheme for proper cut generation. *Technometrics*, **12**, 775–88.

Benzaken, C. (1966). Algorithmes de dualisation d'une fonction booleenne. *Revue Française de Traitement de l'Information Chiffres*, **9**, 119–28.

Benzaken, C. (1968). Structures algébriques de cheminements: pseudo-treillis, gerbiers de carré nul. In *Network and switching theory* (ed. G. Biorci), pp. 40–57. Academic Press, New York.

Boesch, F. T. (1985). The cut frequency vector. In *Graph theory with applications to algorithms and computer science* (ed. Y. Alavi, G. Chartrand, L. Lesniak, D. R. Lick, and C. E. Wall), pp. 103–21. Wiley, New York.

Boesch, F. T. (1986). Synthesis of reliable networks—a survey. *IEEE Transactions on Reliability*, **R-35**, 240–6.

Boesch, F. T. and Chen, S. (1978). A generalization of line connectivity and optimally invulnerable graphs. *SIAM Journal on Applied Mathematics*, **34**, 657–65.

Boesch, F. T., Harary, F., and Kabell, J. A. (1981). Graphs as models of communication network vulnerability: connectivity and persistence. *Networks*, **11**, 57–63.

Boesch, F. T., Satyanarayana, A., and Suffel, C. L. (1990). Some alternate characterizations of reliability domination. *Probability in the Engineering and Informational Sciences*, **4**, 257–76.

Brecht, T. B. and Colbourn, C. J. (1986). Improving reliability bounds in computer networks. *Networks*, **16**, 369–80.

Brecht, T. B. and Colbourn, C. J. (1988). Lower bounds on two-terminal network reliability. *Discrete Applied Mathematics*, **21**, 185–98.

Burt, J. M. and Garman, M. B. (1971). Conditional Monte Carlo: a simulation technique for stochastic network analysis. *Management Science*, **18**, 207–17.

Buzacott, J. A. (1983). The ordering of terms in cut-based recursive disjoint products. *IEEE Transactions on Reliability*, **R-32**, 472–4.
Buzacott, J. A. (1987). Node partition formula for directed graph reliability. *Networks*, **17**, 227–40.
Carré, B. (1971). An algebra for network routing problems. *Journal of the Institute of Mathematics and its Applications*, **7**, 273–94.
Carré, B. (1979). *Graphs and networks*. Oxford University Press, Oxford.
Chari, M. K. and Provan, J. S. (1989). On the use of shellings and partitions in computing reliability. CORS/TIMS/ORSA Meeting, Vancouver.
Chiang, D. T. and Chiang, R.-F. (1986). Relayed communication via consecutive-k-out-of-n:F system. *IEEE Transactions on Reliability*, **R-35**, 65–7.
Chiang, D. T. and Niu, S.-C. (1981). Reliability of consecutive-k-out-of-n:F system. *IEEE Transactions on Reliability*, **R-30**, 87–9.
Chiou, S.-N. and Li, V. O. K. (1986). Reliability analysis of a communication network with multimode components. *IEEE Journal on Selected Areas in Communications*, **SAC-4**, 1156–61.
Chvátal, V. (1973). Tough graphs and hamiltonian circuits. *Discrete Mathematics*, **5**, 215–28.
Cohen, J. E. (1986). Connectivity of finite anisotropic random graphs and directed graphs. *Mathematical Proceedings of the Cambridge Philosophical Society*, **99**, 315–30.
Colbourn, C. J. (1987a). *The combinatorics of network reliability*. Oxford University Press, New York.
Colbourn, C. J. (1987b). Network resilience. *SIAM Journal on Algebraic and Discrete Methods*, **8**, 404–9.
Colbourn, C. J. (1988). Edge-packings of graphs and network reliability. *Discrete Mathematics*, **72**, 49–61.
Dial, R., Glover, F., Karney, D., and Klingman, D. (1979). A computational analysis of alternative algorithms and labeling techniques for finding shortest path trees. *Networks*, **9**, 215–48.
Dodin, B. (1985). Approximating the distribution functions in stochastic networks. *Computers & Operations Research*, **12**, 251–64.
Elmaghraby, S. E. (1967). On the expected duration of PERT type networks. *Management Science*, **13**, 299–306.
Elmaghraby, S. E. (1977). *Activity networks*. Wiley, New York.
Elmaghraby, S. E., Kamburowski, J., and Stallmann, M. F. M. (1989). On the reduction of acyclic digraphs and its applications. OR Report 233. North Carolina State University.
Esary, J. D. and Proschan, F. (1963). Coherent structures of non-identical components. *Technometrics*, **5**, 191–209.
Fishman, G. S. (1986a). A comparison of four Monte Carlo methods for estimating the probability of s-t connectedness. *IEEE Transactions on Reliability*, **R-35**, 145–54.
Fishman, G. S. (1986b). A Monte Carlo sampling plan for estimating network reliability. *Operations Research*, **34**, 581–94.
Fishman, G. S. (1987). A Monte Carlo sampling plan for estimating reliability parameters and related functions. *Networks*, **17**, 169–86.

Floyd, R. W. (1962). Algorithm 97: shortest path. *Communications of the ACM*, **5**, 345.
Fosdick, L. D. and Osterweil, L. J. (1976). Data flow analysis in software reliability. *ACM Computing Surveys*, **8**, 305–30.
Frank, H. and Frisch, I. (1970). Analysis and design of survivable networks. *IEEE Transactions on Communication Technology*, **COM-18**, 501–19.
Frank, H. and Frisch, I. (1971). *Communication, transmission, and transportation*. Addison-Wesley, Reading.
Fulkerson, D. R. (1962). Expected critical path lengths in PERT networks. *Operations Research*, **10**, 808–17.
Gaebler, R. and Chen, R. (1987). An efficient algorithm for enumerating states of a system with multimode unreliable components. Technical Report. U.S. Sprint, Overland Park, Kansas.
Garey, M. R. and Johnson, D. S. (1979). *Computers and intractability: a guide to the theory of NP-completeness*. W. H. Freeman, New York.
Gondran, M. and Minoux, M. (1984). *Graphs and algorithms*. Wiley, Chichester.
Greene, C. (1982). The Möbius function of a partially ordered set. In *Ordered sets* (ed. I. Rival), pp. 555–81. Reidel, Dordrecht.
Hagstrom, J. N. (1988). Computational complexity of PERT problems. *Networks*, **18**, 139–47.
Hagstrom, J. N. (1990). Computing the probability distribution of project duration in a PERT network. *Networks*, **20**, 231–44.
Hagstrom, J. N. and Mak, K.-T. (1986). Computing network reliability with dependent failures. In *Proceedings of the 17th annual conference on modeling and simulation* (ed. W. G. Vogt and M. H. Mickle), pp. 1793–8, Pittsburgh.
Harary, F. (1969). *Graph theory*. Addison-Wesley, Reading.
Hariri, S. and Raghavendra, C. S. (1987). SYSREL: a symbolic reliability algorithm based on path and cutset methods. *IEEE Transactions on Computers*, **C-36**, 1224–32.
Hayhurst, K. J. (1989). A structural factoring approach for analyzing probabilistic networks. Master's thesis. College of William and Mary.
Hayhurst, K. J. and Shier, D. R. (1990). A factoring approach for the stochastic shortest path problem. Technical Report 90-07. Department of Mathematics, College of William and Mary.
Heidtmann, K. D. (1983). Inverting paths and cuts of 2-state systems. *IEEE Transactions on Reliability*, **R-32**, 469–71, 474.
Heidtmann, K. D. (1989). Smaller sums of disjoint products by subproduct inversion. *IEEE Transactions on Reliability*, **38**, 305–11.
Hwang, C. L., Tillman, F. A., and Lee, M. H. (1981). System-reliability evaluation techniques for complex/large systems—a review. *IEEE Transactions on Reliability*, **R-30**, 416–22.
Hwang, F. K. (1982). Fast solutions for consecutive-k-out-of-n:F system. *IEEE Transactions on Reliability*, **R-31**, 447–8.
Hwang, F. K. and Yao, Y. C. (1989). Multistate consecutively-connected systems. *IEEE Transactions on Reliability*, **R-38**, 472–4.
Jensen, P. A. and Bellmore, M. (1969). An algorithm to determine the reliability of a complex system. *IEEE Transactions on Reliability*, **R-18**, 169–74.

Johnson, R. (1984). Network reliability and acyclic orientations. *Networks*, **14**, 489–505.
Karp, R. M. and Luby, M. G. (1983). A new Monte-Carlo method for estimating the failure probability of an n-component system. Report UCB/CSD83/17. Computer Science Division, University of California, Berkeley.
Katona, G. (1966). A theorem of finite sets. In *Theory of graphs, proceedings of Tihany colloquium* (ed. P. Erdös and G. Katona), pp. 209–14. Akademia Kiadó, Budapest.
Kim, Y., Case, K., and Ghare, P. (1972). A method for computing complex system reliability. *IEEE Transactions on Reliability*, **R-21**, 215–19.
Kleene, S. C. (1956). Representation of events in nerve nets and finite automata. In *Automata studies* (ed. C. Shannon and J. McCarthy), pp. 3–40. Princeton University Press, Princeton.
Kleindorfer, G. B. (1971). Bounding distributions for a stochastic acyclic network. *Operations Research*, **19**, 1586–1601.
Kossow, A. and Preuss, W. (1989). Reliability of consecutive-k-out-of-$n:F$ systems with nonidentical component reliabilities. *IEEE Transactions on Reliability*, **38**, 229–33.
Kruskal, J. B. (1963). The number of simplices in a complex. In *Mathematical optimization techniques* (ed. R. Bellman), pp. 251–78. University of California Press, Berkeley.
Lam, Y. F. and Li, V. O. K. (1986a). An improved algorithm for performance analysis of networks with unreliable components. *IEEE Transactions on Communications*, **COM-34**, 496–7.
Lam, Y. F. and Li, V. O. K. (1986b). Reliability modeling and analysis of communication networks with dependent failures. *IEEE Transactions on Communications*, **COM-34**, 82–4.
Li, V. O. K. and Silvester, J. A. (1984). Performance analysis of networks with unreliable components. *IEEE Transactions on Communications*, **COM-32**, 1105–10.
Lipman, M. J. and Pippert, R. E. (1985). Towards a measure of vulnerability II. The ratio of disruption. In *Graph theory with applications to algorithms and computer science* (ed. Y. Alavi, G. Chartrand, L. Lesniak, D. R. Lick, and C. E. Wall), pp. 507–17. Wiley, New York.
Locks, M. O. (1978). Inverting and minimalizing path sets and cut sets. *IEEE Transactions on Reliability*, **R-27**, 107–9.
Locks, M. O. (1982). Recursive disjoint products: a review of three algorithms. *IEEE Transactions on Reliability*, **R-31**, 33–5.
Locks, M. O. (1987). A minimizing algorithm for sum of disjoint products. *IEEE Transactions on Reliability*, **R-36**, 445–53.
Malcolm, D. G., Rosebloom, J. H., Clark, C. E., and Fazar, W. (1959). Application of a technique for research and development program evaluation. *Operations Research*, **7**, 646–69.
Martelli, A. (1974). An application of regular algebra to the enumeration of cut sets in a graph. In *Information processing 74, proceedings of the IFIP congress* (ed. J. L. Rosenfeld), pp. 511–15. North-Holland, Amsterdam.

Martelli, A. (1976). A gaussian elimination algorithm for the enumeration of cut sets in a graph. *Journal of the ACM,* **23,** 58–73.

Martin, J. J. (1965). Distribution of the time through a directed acylic network. *Operations Research,* **13,** 46–66.

McNaughton, R. and Yamada, H. (1960). Regular expressions and state graphs for automata. *IRE Transactions on Electronic Computers,* **9,** 39–47.

Mine, H. (1959). Reliability of physical system. *IRE Transactions on Circuit Theory,* **CT-6,** 138–51.

Minieka, E. and Shier, D. (1973). A note on an algebra for the k best routes in a network. *Journal of the Institute of Mathematics and its Applications,* **11,** 145–9.

Mirchandani, P. B. (1976). Shortest distance and reliability of probabilistic networks. *Computers & Operations Research,* **3,** 347–55.

Moore, E. F. and Shannon, C. E. (1956). Reliable circuits using less reliable relays. *Journal of the Franklin Institute,* **262,** 191–208, 281–97.

Moskowitz, F. (1958). The analysis of redundancy networks. *AIEE Transactions on Communication and Electronics,* **39,** 627–32.

Murchland, J. D. (1965). A new method for finding all elementary paths in a complete directed graph. Report LSE-TNT-22. London School of Economics.

Myrvold, W. (1989). Uniformly most reliable graphs do not always exist. Report DCS-120-IR. Department of Computer Science, University of Victoria.

Nakazawa, H. (1976). Bayesian decomposition method for computing the reliability of an oriented network. *IEEE Transactions on Reliability,* **R-25,** 77–80.

Nel, L. D. and Colbourn, C. J. (1990). Combining Monte Carlo estimates and bounds for network reliability. *Networks,* **20,** 277–98.

Nelson, A. C., Batt, J. R., and Beadles, R. L. (1970). A computer program for approximating system reliability. *IEEE Transactions on Reliability,* **R-19,** 61–5.

Nilsson, N. J. (1980). *Principles of artificial intelligence.* Morgan Kaufmann, Palo Alto.

Page, L. B. and Perry, J. E. (1988). A practical implementation of the factoring theorem for network reliability. *IEEE Transactions on Reliability,* **37,** 259–67.

Page, L. B. and Perry, J. E. (1989). Reliability of directed networks using the factoring theorem. *IEEE Transactions on Reliability,* **38,** 556–62.

Perticone, A. J. (1983). Mathematical approaches to the ranking of athletic teams. In *Proceedings of the 19th annual southeastern TIMS* (ed. J. Eatman), pp. 352–9.

Peyrat, C. (1984). Diameter invulnerability of graphs. *Discrete Applied Mathematics,* **9,** 245–50.

Pippert, R. E. and Lipman, M. J. (1985). Towards a measure of vulnerability I. The edge-connectivity vector. In *Graph theory with applications to algorithms and computer science* (ed. Y. Alavi, G. Chartrand, L. Lesniak, D. R. Lick, and C. E. Wall), pp. 651–7. Wiley, New York.

Politof, T. and Satyanarayana, A. (1986). Efficient algorithms for reliability analysis of planar networks—a survey. *IEEE Transactions on Reliability,* **R-35,** 252–9.

Prékopa, A. (1988). Boole–Bonferroni inequalities and linear programming. *Operations Research*, **36**, 145–62.
Prékopa, A. (1990). Sharp bounds on probabilities using linear programming. *Operations Research*, **38**, 227–39.
Proctor, R. (1982a). Representations of $sl(2, C)$ on posets and the Sperner property. *SIAM Journal on Algebraic and Discrete Methods*, **3**, 275–80.
Proctor, R. (1982b). Solution of two difficult combinatorial problems with linear algebra. *American Mathematical Monthly*, **89**, 721–34.
Provan, J. S. (1986a). Bounds on the reliability of networks. *IEEE Transactions on Reliability*, **R-35**, 260–8.
Provan, J. S. (1986b). Polyhedral combinatorics and network reliability. *Mathematics of Operations Research*, **11**, 36–61.
Provan, J. S. (1986c). The complexity of reliability computations in planar and acyclic graphs. *SIAM Journal on Computing*, **15**, 694–702.
Provan, J. S. and Ball, M. O. (1984). Computing network reliability in time polynomial in the number of cuts. *Operations Research*, **32**, 516–26.
Provan, J. S. and Shier, D. R. (1990). Generating (s, t)-cutsets in directed graphs. Technical Report 90-11. Department of Mathematics, College of William and Mary.
Rai, S. and Aggarwal, K. K. (1980). On complementation of pathsets and cutsets. *IEEE Transactions on Reliability*, **R-29**, 139–40.
Read, R. C. and Tarjan, R. E. (1975). Bounds on backtrack algorithms for listing cycles, paths, and spanning trees. *Networks*, **5**, 237–52.
Resende, L. I. P. (1988). Implementation of a factoring algorithm for reliability evaluation of undirected networks. *IEEE Transactions on Reliability*, **37**, 462–8.
Resende, M. G. C. (1986). A program for reliability evaluation of undirected networks via polygon-to-chain reductions. *IEEE Transactions on Reliability*, **R-35**, 24–9.
Ringeisen, R. D. and Lipman, M. J. (1983). Cohesion and stability in graphs. *Discrete Mathematics*, **46**, 191–8.
Robert, P. and Ferland, J. (1968). Généralisation de l'algorithme de Warshall. *Revue Française d'Informatique et de Recherche Opérationnelle*, **2**, 71–85.
Rosenthal, A. (1975). A computer scientist looks at reliability computations. In *Reliability and fault tree analysis* (ed. R. E. Barlow, J. B. Fussell, and N. D. Singpurwalla), pp. 133–52. SIAM, Philadelphia.
Ross, S. (1985). *Introduction to probability models*. Academic Press, Orlando.
Rota, G.-C. (1964). On the foundations of combinatorial theory I: theory of Möbius functions. *Z. Wahrscheinlichkeitstheorie*, **2**, 340–68.
Roy, B. (1959). Transitivité et connexité. *Comptes Rendus des Séances de l'Académie de Sciences, Paris*, **249**, 216–18.
Satyanarayana, A. (1982). A unified formula for analysis of some network reliability problems. *IEEE Transactions on Reliability*, **R-31**, 23–32.
Satyanarayana, A. and Chang, M. K. (1983). Network reliability and the factoring theorem. *Networks*, **13**, 107–20.
Satyanarayana, A. and Hagstrom, J. N. (1981a). A new algorithm for the

reliability analysis of multi-terminal networks. *IEEE Transactions on Reliability*, **R-30**, 325-34.
Satyanarayana, A. and Hagstrom, J. N. (1981b). Combinatorial properties of directed graphs useful in computing network reliability. *Networks*, **11**, 357-66.
Satyanarayana, A. and Prabhakar, A. (1978). New topological formula and rapid algorithm for reliability analysis of complex networks. *IEEE Transactions on Reliability*, **R-27**, 82-100.
Schwager, S. J. (1984). Bonferroni sometimes loses. *The American Statistician*, **38**, 192-7.
Shanthikumar, J. G. (1982). Recursive algorithm to evaluate the reliability of a consecutive-k-out-of-n:F system. *IEEE Transactions on Reliability*, **R-31**, 442-3.
Shanthikumar, J. G. (1987). Reliability of systems with consecutive minimal cutsets. *IEEE Transactions on Reliability*, **R-36**, 546-50.
Shanthikumar, J. G. (1988). Bounding network reliability using consecutive minimal cutsets. *IEEE Transactions on Reliability*, **R-37**, 45-9.
Shier, D. R. (1973). A decomposition algorithm for optimality problems in tree-structured networks. *Discrete Mathematics*, **6**, 175-89.
Shier, D. R. (1976). Iterative methods for determining the k shortest paths in a network. *Networks*, **6**, 205-29.
Shier, D. R. (1985). Iterative algorithms for calculating network reliability. In *Graph theory with applications to algorithms and computer science* (ed. Y. Alavi, G. Chartrand, L. Lesniak, D. R. Lick, and C. E. Wall), pp. 741-52. Wiley, New York.
Shier, D. R. (1988a). Algebraic aspects of computing network reliability. In *Applications of discrete mathematics* (ed. R. D. Ringeisen and F. S. Roberts), pp. 135-47. SIAM, Philadelphia.
Shier, D. R. (1988b). A new algorithm for performance analysis of communication systems. *IEEE Transactions on Communications*, **36**, 516-19.
Shier, D. R. (1990). Inclusion-exclusion, cancellation, and consecutive sets. Technical Report 90-14. Department of Mathematics, College of William and Mary.
Shier, D. R. and Liu, N. (1989). Bounding the reliability of networks. Technical Report 89-04. Department of Mathematics, College of William and Mary.
Shier, D. R. and Spragins, J. D. (1985). Exact and approximate dependent failure reliability models for telecommunications networks. In *IEEE Proceedings INFOCOM, Washington*, pp. 200-5. IEEE Computer Society.
Shier, D. R. and Whited, D. E. (1985). Algorithms for generating minimal cutsets by inversion. *IEEE Transactions on Reliability*, **R-34**, 314-19.
Shier, D. R. and Whited, D. E. (1986). Iterative algorithms for generating minimal cutsets in directed graphs. *Networks*, **16**, 133-47.
Shier, D. R. and Whited, D. E. (1987). Algebraic methods applied to network reliability problems. *SIAM Journal on Algebraic and Discrete Methods*, **8**, 251-62.
Shier, D. R. and Whited, D. E. (1988). Approximating network reliability. *Congressus Numerantium*, **64**, 221-8.

Shier, D. R., Bibelnieks, E., Jarvis, J. P., and Lakin, R. J. (1990). Algorithms for approximating the performance of multimode systems. In *IEEE Proceedings INFOCOM, San Francisco*, pp. 741-8. IEEE Computer Society.

Shogan, A. W. (1976). Sequential bounding of the reliability of a stochastic network. *Operations Research*, **24**, 1027-44.

Shogan, A. W. (1977a). A recursive algorithm for bounding network reliability. *IEEE Transactions on Reliability*, **R-26**, 322-7.

Shogan, A. W. (1977b). Bounding distributions for a stochastic PERT network. *Networks*, **7**, 359-81.

Spragins, J. D., Sinclair, J. C., Kang, Y. J., and Jafari, H. (1986). Current telecommunication network reliability models: a critical assessment. *IEEE Journal on Selected Areas in Communications*, **SAC-4**, 1168-73.

Stanley, R. (1980). Wehl groups, the hard Lefschetz theorem, and the Sperner property. *SIAM Journal on Algebraic and Discrete Methods*, **1**, 168-84.

Stanley, R. P. (1986). *Enumerative combinatorics*, Vol. 1. Wadsworth & Brooks/Cole, Monterey.

Tarjan, R. E. (1981a). A unified approach to path problems. *Journal of the ACM*, **28**, 577-93.

Tarjan, R. E. (1981b). Fast algorithms for solving path problems. *Journal of the ACM*, **28**, 594-614.

Tsukiyama, S., Shirakawa, I., Ozaki, H., and Ariyoshi, H. (1980). An algorithm to enumerate all cutsets of a graph in linear time per cutset. *Journal of the ACM*, **27**, 619-32.

Valiant, L. (1979). The complexity of enumeration and reliability problems. *SIAM Journal on Computing*, **8**, 410-21.

Valvo, E. J., Shier, D. R., and Jamison, R. E. (1987). Generating the most probable states of a communication system. In *IEEE Proceedings INFOCOM, San Francisco*, pp. 1128-36. IEEE Computer Society.

Van Slyke, R. (1963). Monte Carlo methods and the PERT problem. *Operations Research*, **11**, 839-60.

Van Slyke, R. M. and Frank, H. (1972). Network reliability analysis: Part I. *Networks*, **1**, 279-90.

Warshall, S. (1962). A theorem on boolean matrices. *Journal of the ACM*, **9**, 11-12.

Whited, D. E., Shier, D. R., and Jarvis, J. P. (1990). Reliability computations for planar networks. *ORSA Journal on Computing*, **2**, 46-60.

Wilkov, R. (1972). Analysis and design of reliable computer networks. *IEEE Transactions on Communications*, **COM-20**, 660-78.

Wilson, J. M. (1990). An improved minimizing algorithm for sum of disjoint products. *IEEE Transactions on Reliability*, **39**, 42-5.

Wood, R. K. (1986). Factoring algorithms for computing K-terminal network reliability. *IEEE Transactions on Reliability*, **R-35**, 269-78.

Yang, C.-L. (1989). On generating the most probable states of networks with multimode components. Technical Note. GTE Laboratories, Waltham, Massachusetts.

Yang, C.-L. and Kubat, P. (1989). Efficient computation of most probable states for communication networks with multimode components. *IEEE Transactions on Communications*, **37**, 535–8.

Yoo, Y. B. and Deo, N. (1988). A comparison of algorithms for terminal-pair reliability. *IEEE Transactions on Reliability*, **R-37**, 210–15.

Zemel, E. (1982). Polynomial algorithms for estimating network reliability. *Networks*, **12**, 439–52.

Zimmermann, U. (1981). *Linear and combinatorial optimization in ordered algebraic structures*, Vol. 10. Annals of Discrete Mathematics. North-Holland, Amsterdam.

Index

adjacency matrix 27, 31, 55, 66
antichain 109–10

bipartite 20, 90
bounds
  algebraic 48
  Bonferroni 42
  Chari–Provan 48
  disjoint-cutsets 46–7
  disjoint-paths 46–7
  Esary–Proschan 44–5
  Kruskal–Katona 48

cancellation 9, 11, 82, 100
chain 75, 80–2, 106
coherent system 60, 74, 83, 87–9, 91, 110, 113
cohesion 4
contraction 12, 16–17
convex set 75, 92, 108
convolution 122–5
critical path 118, 129
critical set 110
cut 67–9, 73, 79
cut frequency vector 4
cutset 9, 20, 61, 73, 82–3
  minimum cardinality 19
cutset enumeration 45, 67

dependence
  edge 22–3, 48
  path 119
disjoint products 13, 21, 83
domination 17, 20

edge connectivity 3–4
edge cover 88–90

factoring 10, 15, 20–1, 125–7, 129
factoring edge 124–5
failure set 87

Gauss–Seidel method 31–2
Gauss–Jordan method 66–7, 70–1
generating function 110
  edge 120
  path length 121

Hasse diagram 73, 104–6

inclusion–exclusion 10, 20–1, 82–3, 100
integrity 5
inversion 71
irrelevant edge 12

Jacobi method 30–2, 49

k-out-of-n system 75–6, 98–100

language 64–5
lattice 72
  distributive 23, 106
  height 106, 110
  rank 106, 109–10
LU decomposition 55

matroid 14
mincut 74, 83, 87–9
minimal solution 29, 55
minpath 74, 83, 87, 89
Möbius inversion 80, 83
most probable state 101–2, 106, 114
multistate system 113, 118

network 2
  acyclic 12, 28–9, 31, 47, 52, 127–9
  PERT 129
  planar 20, 71, 74
  series-parallel 15, 17, 47, 122
node connectivity 4
NP-complete 18

#P-complete 19–21, 52, 88–90
partially ordered set 73, 80, 104, 121
path enumeration 20, 62, 71
perfect matching 89–90
performance measure 3, 84, 101
persistence 5
pivotal decomposition 10, 12, 15
pseudopolynomial algorithm 72

reductions
  conditioned 124
  fundamental 122
  parallel 15–18, 23, 120, 122
  series 15–18, 23, 120, 122
  two-neighbour 15, 123
redundancy 6
reliability
  all-terminal 5, 14, 19, 21, 60–1
  covering 84
  K-terminal 5, 13, 18, 21, 83
  polynomial 22, 24, 27
  two-terminal 5, 8
resilience 6

semilattice 75–7, 81, 91
shellable system 14, 60
shortest path 29, 48, 59
  stochastic 118
Sperner property 109
state space 113
  approximation to 101
  enumeration of 8
structure function 74, 110
success set 87
system reliability 18, 74, 88

threshold system 14
topological formula 13, 82, 100
topological numbering 52, 77, 81, 128–9
  reverse 97
toughness 5
two-terminal unreliability 10, 19, 42

vulnerability 3–5, 7